KU-737-483

Contents

933

FOETUS INTO MAN

WITHDRAWN

Inst. of Biological Anthropology
58 Banbury Road, Oxford, OX2 6QS

Tel: 0865 - 274700   Fax: 0865 - 274699

This book was kindly donated to the
Biological Anthropology Library by
Professor G. Ainsworth Harrison
on his retirement in 1994

WITHDRAWN

Foetus into Man:
Physical Growth from
Conception to Maturity

J. M. Tanner

WITHDRAWN

Inst. of Biological Anthropology
58 Banbury Road, Oxford OX2 6QS

Tel: 0865 - 274700   Fax: 0865 - 27469

Open Books
London

THE TYLOR LIBRARY

WITHDRAWN

UNIVERSITY OF OXFORD

First published in 1978 by Open Books Publishing Ltd,
11 Goodwin's Court, London WC2N 4LB

© J. M. Tanner 1978

Hardback ISBN 0 7291 0089 8
Paperback ISBN 0 7291 0084 7

This title is available in both hardback and paperback
editions. The paperback edition is sold subject to the
condition that it shall not, by way of trade or otherwise,
be lent, re-sold, hired out or otherwise circulated in
any form of binding or cover other than that in which it
is published and without a similar condition, including
this condition, being imposed on the subsequent purchaser.

All rights reserved. No part of the publication may be
reproduced or transmitted in any form or by any means,
electronic or mechanical, including photocopy, recording,
or any information storage and retrieval system, without
permission in writing from the publisher.

*To Reg Whitehouse, my friend and collaborator for
twenty-nine years, whose contributions to human
auxology are implicit in all these pages*

Printed in Great Britain by
Fletcher & Son Ltd, Norwich

Introduction

This is a book which describes the process of growth in children in terms which I hope the biologically unsophisticated reader will understand and the biologically sophisticated approve. Where complexity exists, I have said so plainly, avoiding, I hope, all facile generalizations. I have sometimes entered into details, but only when they are elegant, instructive and undisputed. Biology is like a Van Eyck painting; as you approach closer, clarity actually increases. The focus holds true down to the tiniest detail. The clothes of the street vendor seen in the far distance through the window are as clear, under the magnifying-glass, as the robes of the foreground personages talking in the room. Such perspective is not of the eye but of the mind; and I must beg the reader to use an equal facility of focus as we traverse between broad and narrow views.

Though a little biological knowledge on the reader's part is certainly helpful, I have not assumed it. Chromosomes, endocrines, proteins, codons, centiles and regressions are all explained as we go along. I have, however, assumed that readers know the names of parts of the body, including those of which they have no personal experience.

Chapter 1 presents the phenomenon we are going to study, the curve of growth of the child. It is the curve of height that is given; this is the characteristic we single out in everyday speech, for we talk of a child 'growing up'. But at the outset one must realize that growth is far from being a simple and uniform process of becoming taller or larger. As the child gets bigger there are changes in shape and in body composition and in the distribution of various tissues. In the newborn the head represents about a quarter of the total length; in the adult about one-seventh. In the newborn much of the muscle still consists of water; in the three-year-old the muscle cells are packed with contractile protein. Different tissues (such as muscles, nerves, liver) mature at different rates and the growth of a child consists of a highly complex series of changes. It is like the weaving of a cloth whose design never repeats itself. The underlying threads, each coming from its reel at its own rhythm, interact with one another continuously, in a manner always highly regulated and controlled.

The fundamental biological questions of growth relate to these processes of regulation, to the programme that controls the loom. At present very little about this is understood. In Chapter 2 we discuss cells, the fundamental units of growth, and distinguish between organs such as the skin, where cells are continuously produced to replace those that continuously die, and organs such as the nerves and muscles where cells, once formed, remain throughout most of the animal's life, and no new cells can be formed once the growth period is over. There is too little certainty about the way that cell and organ growth is regulated for a long consideration of this to be included, but in Chapter 10 we return to consider the regulation of growth in the whole organism, after we have accumulated sufficient knowledge to make some sense of it. Perhaps it is worth saying to the non-biologist reader that if Chapter 2 proves difficult for him, which I hope it will not, he can take courage from the fact that all the rest is easier.

In Chapter 3, prenatal growth is described. Perhaps the most surprising fact that has emerged in the last few years is how many fertilized ova fail to result in a live baby. About 50% of all conceptuses are spontaneously aborted, nearly all because of intrinsic faults in their make-up. We have to adjust ourselves to the notion that children with certain diseases such as Down's syndrome (or mongolism) and Turner's syndrome are the rare survivors from cohorts nearly all of whose members die in the first few weeks after conception.

In Chapter 4, the chromosomal and endocrine control of sex differentiation in utero and in early childhood is described, and in Chapter 5 the further differentiation that takes place at puberty. The last two decades have seen a great increase in our understanding both of the bodily changes of puberty (and especially of their great variability), and of the endocrine mechanisms which control them. The whole subject of childhood and pubertal endocrinology is at last beginning to make sense, owing chiefly to our growing ability to measure very small amounts of hormones by chemical or biological means. Further advances will now depend increasingly on studies of normal adolescents followed individually all through puberty, and that demands another and perhaps more difficult skill than the manipulation of chemical assays.

Different children experience puberty at very different ages, so that amongst a group of twelve-year-old girls or fourteen-year-old boys there will be some who have not yet started their pubertal changes and others who have practically finished them. This variability, which is almost entirely biological in origin, has important social and educational consequences. Though seen most clearly at puberty, the differences in tempo of growth – whether the course of growth is played out quickly or slowly – are present at all ages. The concept of physiological maturity, or developmental age, has thus arisen, and Chapter 6 is devoted to its description, measurement and implications.

In Chapters 7 and 8 two of the most important organ systems have been singled out: the endocrine glands because they play such an overwhelmingly important part in the control of growth and sex development, and the brain because it is of such interest and importance to human beings. Our increased understanding of endocrine events has enabled one particularly spectacular advance in medicine to be made. Some children – mostly boys – grow up to be very small, though of quite normal proportions, because they lack the growth hormone normally secreted by the pituitary gland. Normal growth is restored by injection of the growth hormone (just as injections of insulin restore the metabolic situation of diabetics). Only growth hormone from humans works, however, so it has to be obtained from human pituitaries after death. This is now done on a large scale in many countries, and these children if diagnosed early enough can look forward to becoming perfectly normal-sized

No such spectacular advance has taken place regarding brain development. However, there is now quite a large number of research workers throughout the world devoting themselves to this technically very difficult field, and it seems safe to predict that in the next ten years we shall have achieved real knowledge about how the complex structure of the brain is generated during growth and even about the ways memories are stored. Already the work of Hubel and Wiesel and their many followers, described in Chapter 8, shows how the visual system, including the cells of the cerebral cortex, depends for its development on a precise interaction between the environment and the nerve cells. This interaction has to occur at a particular time, known as a 'sensitive period', in the animal's maturational calendar. (Sensitive periods pervade the whole of growth, as the reader will soon discover.)

Chapter 9 is a long one. It discusses the effects of the interaction of genes and environment on growth. Differences in growth be-

tween one family and another, and between one population group
and another, are detailed; the effects of undernutrition are described, and of factors associated with social class, numbers of
brothers and sisters, climate, psychological stress, illness and
urbanization. In this field also things are rapidly becoming clearer.
We are beginning to have examples of populations of all the major
racial groups living under similar environmental circumstances. For
example, in the U.S.A. persons of European and of African origin
growing up under similar circumstances have very similar growth
curves for height though not for body proportions. Under similar
conditions Asians from Japan and China seem to be a little smaller
and to mature a little faster. Throughout the industrialized world
there has been a tendency during the last hundred years for children
to become larger and reach maturity earlier, a tendency now reaching its end. Though rate of growth remains one of the most useful
of all indices of public health and economic well-being in developing and heterogeneously developed countries, it must not be
thought that bigger, or faster, is necessarily better. From an
ecological point of view smallness has advantages; here also a
change of attitude has recently been developing, with simplicity
giving way to a more sophisticated search for balance.

In Chapters 11 and 12 standards for normal growth are given
and some of the common pathologies of growth are briefly described. The standards provide a do-it-yourself health visitor's kit
for parents interested in seeing how their children grow (comparing
children's growth curves by marks made on a door specially designated for this purpose was a common pastime in large Victorian
families). The proper method of measuring length or height (all too
seldom followed by the health professionals) is described; how to
plot the results on a chart; and how, at certain ages, to see if the
child takes after the parents. Tables for predicting the probable
adult height are also given. In all these standards there is very great
normal variation and readers must be warned to read exactly and
with caution. A famous statistician once observed that, unfortunately, there was no way of preventing half the people being below
the average. Similarly, 3% of perfectly normal children are below
the line on the chart called the 3rd centile, which is often taken as
the arbitrary limit of normality. However, if a child is far below the
standards then certainly the cause for this ought to be sought,
particularly nowadays when growth-hormone deficiency, at least,
can be successfully treated.

Finally, as to further reading: it seemed wiser in general not to put references in the text, where the general reader may find them distracting. I have provided instead a two-tiered bibliography for each chapter, consisting of a short list of further reading in which each paper is annotated, and a longer list of references without annotations. Where authors' names are mentioned in the text the relevant papers will be found listed in the References.

Very many colleagues and friends have helped by criticizing early drafts. I am especially grateful for the comments of Bryan Senior, Vreli Fry, Marjorie Baines and Kenneth Sass, who undertook the arduous task of reducing to a minimum the number of paragraphs incomprehensible or boring to the non-biologist. Messrs R. H. Whitehouse and Noel Cameron, and Professor Jim Garlick and his students at Cambridge, served as human biologist assessors, and Dr Michael Preece, Dr B. A. Tanner and Professor W. A. Marshall as clinical scrutineers. The faults that remain are mine. Lastly, I wish to thank Ray Lunnon and Brian Kesterton for the pictures and diagrams, Jan Baines for assembling the material and dissembling, at times, my presence in the office, and Sara Dunford for her cheerfulness and care over innumerable typescripts.

J. M. Tanner
Institute of Child Health, London, July 1977

CHAPTER 1
The Curve of Growth

Figure 1 shows the most famous of all records of human growth. It concerns the height of a single boy, measured every 6 months from birth to 18 years. This is the oldest longitudinal record in existence, and it remains, for our purpose of illustration, one of the best. It was made during the years 1759–77 by Count Philibert Guéneau de Montbeillard on his son and it was published by Buffon in a supplement to the *Histoire Naturelle*.

In Figure 1a is plotted the height attained at successive ages; in 1b the increments in height from one age to the next, expressed as the rate of growth per year. If we think of growth as a form of motion (analogous to the journey of a train) then the upper curve is one of distance travelled; the lower curve, one of velocity. The velocity, or rate of growth, naturally reflects the child's state at any particular time better than does the distance achieved, which depends largely on how much the child has grown in all the preceding years. Thus, for those substances which change in amount with age, the concentrations in blood and tissues are more likely to run parallel to the velocity than to the distance curve. Indeed, in some circumstances it is the acceleration rather than the velocity curve which best reflects physiological events.

Figure 1b shows that in general the velocity of growth decreases from birth (and actually from about the fourth month of foetal life; see Chapter 3), but that this decrease is interrupted shortly before the end of the growth period. At this time, from 13 to 15 years in this particular boy, there is a marked acceleration of growth, called the adolescent growth spurt. (Some writers distinguish sharply the terms 'adolescence' and 'puberty' though not all who do so agree in their distinctions. Some use puberty to refer to physical changes, adolescence to refer to psychosocial ones. I have used the terms interchangeably in this book.) From birth until age 4 or 5 the rate of growth in height declines rapidly, and then the decline, or deceleration, gets gradually less, so that in some children the velocity is practically constant from 5 or 6 up to the beginning of the adolescent spurt. A slight increase in velocity is sometimes said to occur between about 6 and 8 years, providing a second wave on

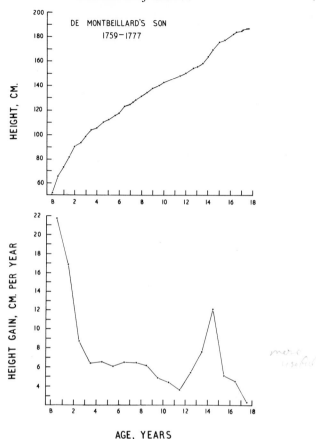

FIG. 1. Growth in height of de Montbeillard's son from birth to 18 years (1759–77): (*above*) distance curve, height attained at each age; (*below*) velocity curve, increments in height from year to year (From Tanner, 1962, drawn from data of Scammon, 1927)

the general velocity curve. Although Figure 1 seems to show its presence, examination of many other individual records from ages 3 to 13 fails to reveal it in the great majority; if it occurs at all, it is in a minority of children.

Growth is in general a very regular process. Contrary to opinions still sometimes met, growth in height does not proceed by stops and starts, nor does growth in upward dimensions alternate with growth in transverse or, more ominously, anteroposterior ones. The more carefully the measurements are taken, with precautions, for example, to minimize the decrease in height that occurs during the

Foetus into Man

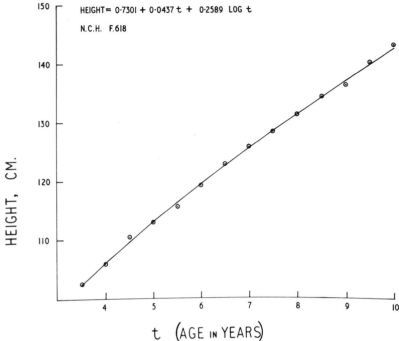

FIG. 2. Curve of form $y = a + bt + c \log t$ fitted to stature measurements taken every 6 months by R. H. Whitehouse on a girl from ages 3·5 to 10·0 years
(Harpenden Growth Study)
(From Israelsohn, 1960)

day for postural reasons, the more regular does the succession of points in the graph become. Figure 2 shows the fit of a smooth mathematical curve to a series of measurements taken on a child by the same observer, my colleague R. H. Whitehouse, every 6 months from ages 3·5 to 10·0. None of the points deviates from the line by an amount more than measuring-error. This is generally true, although in some children regular seasonal variations (discussed in Chapter 9) superimpose an added 6-month rhythm about the curve. There is no evidence for 'stages' of growth in height (or any other physical measurement) except for the spurt associated with adolescence. Perhaps increments of growth at the cellular level are discontinuous; but at the level of bodily measurements, even of single bones measured by x-rays, we can only discern complete continuity, with a velocity that changes gradually from one age to the next.

Many attempts have been made to find mathematical curves which fit, and thus summarize, human and animal growth data. Most have ended in disillusion or fantasy; disillusion because fresh data failed to conform to them, fantasy because the system eventually contained so many parameters (or 'constants') that it became impossible to interpret them biologically. What is needed is a curve or curves with relatively few constants, each capable of being interpreted in a biologically meaningful way; yet the fit to actual data must be adequate, within the limits of measuring-error. Part of the difficulty arises because the measurements usually taken are themselves biologically complex. Stature, for example, consists of leg length, trunk length and head height, all of which have rather different growth curves. Even with such relatively homogeneous measurements as forearm length or calf–muscle width, it is not clear what purely biological assumptions should be made as the basis for the form of the curve. The assumption that cells are continuously dividing leads to a different formulation from the assumption that cells are adding constant amounts of non-dividing material, or adding amounts of material at rates varying from one age period to another.

But fitting a curve to the individual values is the only way of extracting the maximum information about an individual's growth from the measurement data. This fact becomes increasingly inescapable when the effect of environmental circumstances on growth rate (e.g. illness on height growth) is investigated, or when two different measurements are being compared for the consistency of each as the child grows up. The individual's consistency can only be measured by deviations from his own growth curve. A change of rank order of two individuals in a measurement from one age to another may represent not inconsistency but consistently differing rates of change, one individual having a small velocity in the measurement and the other individual a larger one.

Very recently my colleague Dr M. Preece, in collaboration with Dr M. Baines of Reading University, has succeeded in formulating a curve which fits data on height and its components from soon after birth right up to maturity. The curve is at first the shape shown in Figure 2, but becomes s-shaped at adolescence. In its simpler form it has only four parameters (or constants) so that all the measurements of de Montbeillard's son, for example, can be subsumed in a set of four numbers. Previously it was necessary to use two curves, one up till and the other throughout adolescence.

Types of Growth Data

Such curves have to be fitted to data on single individuals. Curves
of the yearly averages derived from different children each
measured once only in general have a different shape. Thus the
distinction between the two sorts of investigation is very important.
The method of study using the same child at each age is called
longitudinal; that using different children at each age is called *cross-
sectional*. In a cross-sectional study each child is measured only
once and all the children at age 10, for example, are different from
those at age 9. A longitudinal study may extend over any number
of years; there are short-term longitudinal studies extending from
ages 3 to 5, for instance, and full birth-to-maturity longitudinal
studies in which the children may be examined once, twice, or even
more times every year from birth until 20 or over.

In practice it is always impossible to measure exactly the same
group of children every year for a prolonged period; inevitably
some children leave the study, and others, if that is desired, join it.
A study in which this happens is called a *mixed longitudinal* study,
and special statistical techniques are needed to get the maximum
information out of its data. One particular type of mixed study is
that in which a number of relatively short-term longitudinal groups
are interlocked; for example, we may simultaneously study four
groups, one followed from 0 to 6 years, the second from 5 to 11, the
third 10 to 16, and the fourth 15 to 20. Thus the whole age-range is
covered for estimates of mean yearly velocity in the research time
of five years. However, problems arise at the 'joins' unless the
sampling has been remarkably good.

Both cross-sectional and longitudinal studies have their uses, but
they do not give the same information and cannot be dealt with in
the same way. Cross-sectional surveys are obviously cheaper and
more quickly done, and can include far larger numbers of children.
They tell us a good deal about the curve of growth of height or
weight for a given age (the 'distance' curve). It is essential to have
cross-sectional surveys as part-basis for constructing standards for
height and weight and other measurements in a given community,
and periodic surveys are valuable in assessing the nutritional
progress of a country or of particular socio-economic groups, or
the health of the child population as a whole. But cross-sectional
curveys have one great drawback; they can never reveal individual
differences in rate or velocity of growth or in the timing of particular phases such as the adolescent growth spurt. It is these

individual differences in growth velocity which chiefly throw light on the genetical control of growth and on the correlation of growth with psychological development, educational achievement and social behaviour. Longitudinal studies are laborious and time-consuming; they demand great perseverance on the part of those who make them and those who take part in them; and they demand very high technical standards, since, in the calculation of a growth increment from one occasion to the next, two errors of measurement occur. They demand also a sequential approach to problems, with all past data fully computerized and available for analysis in relation to a specific question at any time. The evidence of the past suggests that unless accompanied by cross-sectional surveys and animal experimentation, as they are, for example, at the Fels Research Institute in Yellow Springs, Ohio and at the Department of Growth and Development of the Institute of Child Health of London University, they can sink over the years into sterile deserts of number-collecting. But they are indispensable.

Cross-sectional data can be in some important respects misleading. Figure 3 illustrates the effect on 'average' figures produced by the individual differences in the age at which the adolescent spurt begins. Figure 3a shows a series of individual velocity curves from 10 to 18 years, each individual starting his spurt at a different age. The average of these curves, obtained simply by treating the values cross-sectionally and adding them up at ages 10, 11, 12, etc. and dividing by 5, is shown by the dashed line. This line in no way characterizes the 'average' velocity curve; on the contrary, it is a travesty of it. It smoothes out the adolescent spurt, spreading it along the time axis. It does not take account of the 'phase-differences' between the individual curves. Figure 3b shows the same individual curves, but arranged so that their peak velocities coincide; the average curve then characterizes the group in a proper manner. In passing from Figure 3a to 3b the time scale has been altered so that in 3b the curves are plotted not against chronological age but against a measure which arranges the children according to how far they have travelled along their course of development; in other words, they are arranged according to their true developmental or physiological growth-status. This is nearly always the appropriate method in analysing longitudinal data, especially at adolescence.

Averages computed from cross-sectional data, however, inevitably produce velocity curves of this flattened, distorted type;

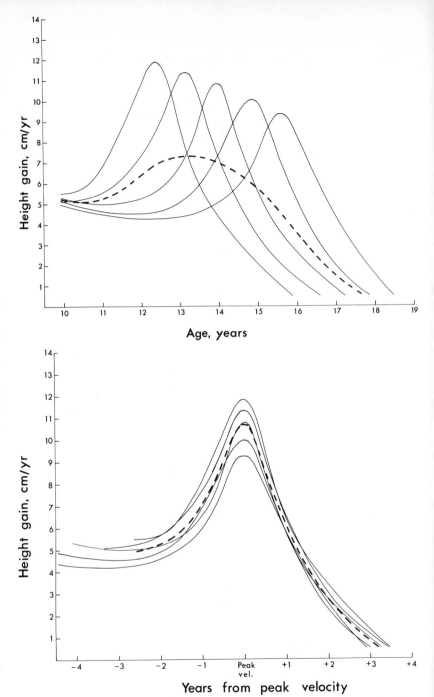

FIG. 3. The relation between individual and mean velocities during the adolescent spurt: (*above*) the individual height velocity curves of five boys of the Harpenden Growth Study (solid lines) with the mean curve (dashed) constructed by averaging their values at each age; (*below*) the same curves all plotted according to their peak height velocity

(From Tanner, Whitehouse and Takaishi, 1966)

and, equally, distance curves show the distortion by not rising sufficiently rapidly at adolescence.

Until recently all the published height and weight standards used
in hospitals and schools incorporated this distortion. However, it is
possible to construct a curve which represents the actual growth of
a typical individual, by taking the shape of the curve from
individual longitudinal data, and the absolute values for the beginning and end from large cross-sectional surveys. Figures 4 and 5
show height-attained and height velocity curves for the 'typical' boy
and girl in Britain in 1965, determined in this way. By 'typical' is
meant a boy or girl who has the mean birth length, grows always at
the mean velocity, has the peak of the adolescent growth spurt at
the mean age, and, finally, reaches the mean adult height at the
mean age of cessation of growth. There is, of course, a certain
danger in showing such a smooth, average curve, for measurements
on a single individual are naturally less regular, however expert the
measurer. Almost no individual exactly follows the curves of

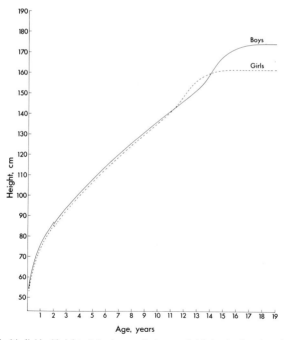

Fig. 4. Typical-individual height-attained curves for boys and girls (supine length to the age of 2;
integrated curves of Fig. 5)
(From Tanner, Whitehouse and Takaishi, 1966)

Figures 4 and 5; but most individuals have growth curves of this shape. Standards for height, constructed around these curves, are given in Chapter 11.

Boys' and Girls' Height Curves

Figures 4 and 5 show the height curves from birth to maturity. Up to age 2 the child is measured lying down on his back. One examiner holds his head in contact with a fixed board, and a second person stretches him out to his maximum length and then brings a moving board into contact with his heels (see Figure 58, p. 174). This measurement is called supine length, and averages about 1 cm. more than the measurement of standing height taken on the same child, even when, as in the best techniques, the child is urged to stretch upwards to full height and is aided in doing so by a meas-

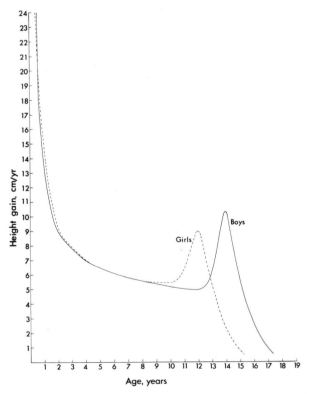

FIG. 5. Typical-individual velocity curves for supine length or height in boys and girls. These curves represent the velocity of the typical boy and girl at any given instant.
(Modified from Tanner, Whitehouse and Takaishi, 1966)

urer applying gentle upward pressure to the bony prominences behind the ears (Figure 57, p. 174). This difference in method of measurement causes the break in the line in Figure 4 at age 2.

Figure 4 shows the typical girl as slightly shorter than the typical boy at all ages until adolescence. She becomes taller shortly after 11·0 years because her adolescent spurt takes place two years earlier than the boy's. At age 14·0 she is surpassed again in height by the typical boy, whose adolescent spurt has now started, whereas hers is nearly finished. In the same way, the typical girl weighs a little less than the boy at birth, equals him at age 8·0, becomes heavier at age 9 or 10 and remains so till about age 14·5.

The velocity curves given in Figure 5 show these processes more clearly. At birth the typical boy is growing slightly faster than the typical girl but the velocities become equal at about 7 months and then the girl grows faster until about age 4·0. From then till adolescence no difference in velocity can be detected. The sex difference is best thought of, perhaps, in terms of acceleration, the boy decelerating harder than the girl over the first four years. The typical girl begins her adolescent height spurt at about 10·5 and reaches peak height velocity at approximately 12·0 in the U.K. and about three months earlier in the U.S.A. The boy begins his spurt and reaches his peak just two years later. The boys' peak is higher than the girls', averaging in our data 10·3 centimetres a year compared with the girls' 9·0 cm/yr (as 'instantaneous' peaks, i.e. peaks obtained by fitting smooth curves to the observations; the velocities over the whole year which includes the peak moment are naturally less, averaging 9·5 cm/yr for boys and 8·1 cm/yr for girls).

Girls are always in advance of boys (i.e. closer to their final mature status), even at birth; this very important sex difference is considered in more detail in Chapters 5 and 6.

Growth Curves of Different Tissues and Different Parts of the Body
Most body measurements follow approximately the growth curves described for height. The great majority of skeletal and muscular dimensions grow in this manner, and so also do the internal organs such as liver, spleen and kidney. But some exceptions exist, most notably the brain and skull, the reproductive organs, the lymphoid tissues of the tonsils, adenoids and intestines, and the subcutaneous fat.

In Figure 6 these differences are shown, using the size attained by various tissues as a percentage of the birth-to-maturity increment.

Foetus into Man

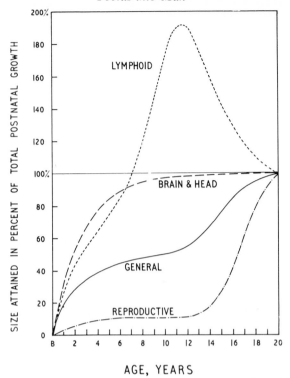

Fig. 6. Growth curves of different parts and tissues of the body, showing the four chief types. All the curves are of size attained, and plotted as percentage of total gain from birth to 20 years, so that size at age 20 is 100 on the vertical scale.

Lymphoid type: thymus, lymph nodes, intestinal lymph masses

Brain and head type: brain and its parts, dura, spinal cord, optic apparatus, cranial dimensions

General type: body as a whole, external dimensions (except head), respiratory and digestive organs, kidneys, aortic and pulmonary trunks, musculature, blood volume

Reproductive type: testis, ovary, epididymis, prostate, seminal vesicles, fallopian tubes

(From Tanner, 1962, redrawn from Scammon, 1930)

Height follows the 'general' curve. The reproductive organs, internal and external, have a slow pre-pubescent growth, followed by a very large adolescent spurt.

The brain, together with the skull covering it and the eyes and ears, develops earlier than any other part of the body and thus has a characteristic postnatal curve. (Brain growth is further discussed in Chapter 8.) If the brain has any adolescent spurt at all, it is a very small one. A small but definite spurt occurs in head length and breadth, but most or all of this is due to thickening of the skull bones and the scalp, and development of the air sinuses. The dimensions of the face follow a path somewhat closer to the general curve. There is a considerable adolescent spurt, especially in the mandible, resulting in the jaw's becoming longer and more project-ing, the profile straighter, and the chin more pointed. But, as always in growth, there are considerable individual differences, to the degree that a few children have no detectable spurt at all in some face measurements.

The eye probably has a slight adolescent spurt, although present data are not sufficiently accurate to make us certain. Very likely it is this that is responsible for the increase in frequency of short-sightedness which occurs at the time of puberty. Although myopia increases continuously from at least age 6 to maturity, a par-ticularly rapid rate of change occurs at about 11 to 12 in girls and 13 to 14 in boys, as would be expected if there was a rather greater spurt in the axial dimension of the eye than in its vertical dimen-sion.

The lymphoid tissue has quite a different growth curve from the rest of the body. It reaches its maximum amount before adoles-cence and then, probably under the direct influence of sex hor-mones, declines to its adult value.

The subcutaneous fat layer also has a curve of its own, of a slightly complicated sort. Its thickness can be measured either by x-rays or, more simply, at certain sites in the body, by picking up a fold of skin and fat between the thumb and forefinger and measur-ing its thickness with a special, constant-pressure caliper. Figure 7 shows the distance curves of skinfolds taken half-way down the back of the arm (triceps) and at the back of the chest, just below the shoulder blade (subscapular). Subcutaneous fat begins to be laid down in the foetus at about 34 weeks postmenstrual age (see p. 41), and increases continuously, to reach a peak at about 9 months after birth. (This is in the average child; the peak may be reached as early as 6 months or as late as 12 or 15.) After 9 months, the skinfolds decrease until age 6 to 8 when they begin to rise again. Girls have a little more total fat than boys at birth, and the difference becomes gradually more marked during childhood. From 8 years on, the curves for girls and boys diverge more radically, as do the curves for limb and body fat. At adolescence the limb fat in boys on average decreases (see triceps, Figure 7); the body fat (subscapular) shows a temporary slowing down of gain, but no loss. In girls there is a slight halting of the limb-fat gain at adolescence, but no loss; the trunk-fat shows only a steady rise until adulthood. The range of

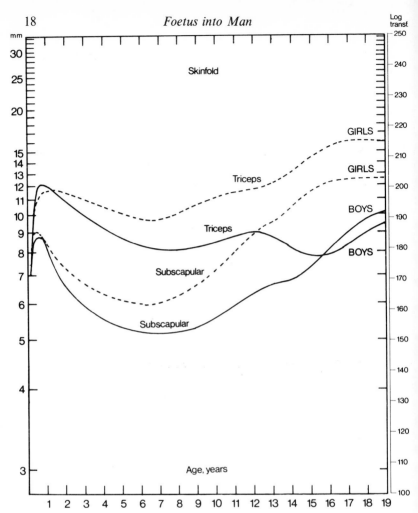

Fig. 7. Distance curve of subcutaneous tissue measured by Harpenden skinfold calipers over triceps (back of upper arm) and under scapula (shoulder blade). Scale is mm. on the left and logarithmic transformation units on he right-hand side. British children, 50th centiles (From Tanner and Whitehouse, 1975)

skinfold thicknesses at each age from birth onwards in English children is given in Tanner and Whitehouse (1975), and in United States schoolchildren in Johnston *et al.* (1974).

Postadolescent Growth

Growth, even of the skeleton, does not entirely cease at the end of the adolescent period. The limb bones stop increasing in length, but the vertebral column continues to grow until about age 30, by apposition of bone to the tops and bottoms of the vertebral bodies. Thus height increases by a small amount, on average 3 to 5 millimetres. From about 30 to 45 years height remains stationary, and then it begins to decline. Head length and breadth and facial diameters increase slightly throughout life; and so do the widths of the bones in the leg and in the hand, in both sexes. For practical purposes, however, it is useful to have an age at which we may say that growth in stature virtually ceases, i.e. after which only some 2% is added. At present in North America and north-west Europe the average boy stops growing, in this sense, at 17·5 years and the average girl at 15·5 years. There is a normal range of variation amongst individuals, amounting to about two years, on either side of these averages.

Individual Variation in Growth

There is very wide variation amongst children both in height at any age and in the velocity of height growth from one age to the next. The extent of this variation is described in some detail in Chapter 11, where charts enabling readers to place themselves and their children in relation to others are given. It is necessary here, however, to give a brief description of the way in which individual variation is measured, since we shall need to have some of the terminology in mind during many of the succeeding chapters.

Figure 8 shows the standards for height at successive ages for boys in the U.K. Each of the lines represents a *centile* (or percentile). Their meaning is as follows. Imagine 100 boys all aged 9·0 years. Measure them and line them up precisely in order of increasing height. Then place a line marked 3 so that 3 out of the 100 lie just below it. Place further lines such that 10 boys lie below the 10th centile line, 25 below the 25th centile and just half of the boys below the 50th centile line. At the top end 97 boys lie below the 97th centile (i.e. only 3 boys lie above it). Notice that this system can be applied whatever the shape of the curve traced out by the heads of those 100 boys lined up in order of increasing height. If this line of heads rose at first gently, then much more suddenly at the higher values, the centile approach would still be valid.

Readers who are familiar with statistics (and those to whom statistics is a closed book will miss nothing vital by skipping the remainder of this paragraph) will know that the 50th centile is the same as the mean in the Normal or Gaussian distribution. Height is

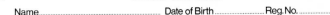

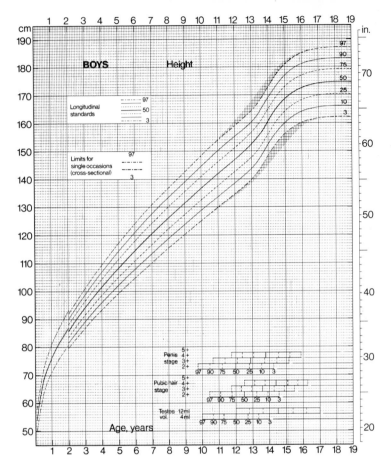

FIG. 8. Centile standards for boy's height, birth to 19 years
British data
(From Tanner and Whitehouse, 1976)

in fact distributed in this way and it is for this reason that conventionally the outside centiles are taken as the 3rd and 97th, which are rounded forms of the 2·5th and 97·5th, corresponding respectively to plus and minus 2 standard deviations (analogous to the 5% probability level in a two-sided significance test). But weight is not distributed Normally at most ages, and nor are skinfolds. For these distributions the centile positions are still valid while standard deviation scores are not. The accuracy with which the centiles are located, however, depends on the distribution, the standard error of the centile being greater the more the distribution departs from Normal.

Conventionally we say that a child below the 3rd centile should be regarded as possibly pathological. It must be stressed that this is a convention and nothing more; there is no magic in the number 3, at least in this context. If the next stop in determining whether a child was abnormal was something drastic like an exploratory operation, then clearly we would demand a much lower level than the 3rd or even the 1st centile, to avoid operating on children who were normal. If, on the other hand, it only meant giving extra morning milk and biscuits, we might choose the 10th centile to make sure we got all the undernourished within our treatment net. However, if we set up very stringent standards, like the 1st centile, we run the risk of leaving some truly pathological children in amongst the healthy ones. We shall see further how the theory of standards works in practice in Chapter 11.

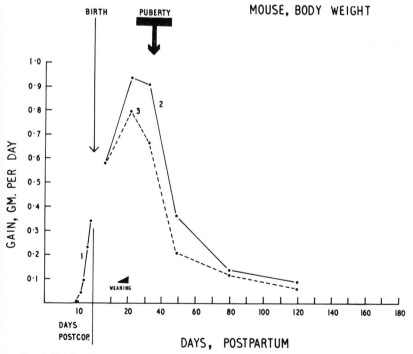

FIG. 9. Weight velocity curve for the mouse: curve 1, sexes combined, cross-sectional; curves 2 and 3, males (18) and females (18), pure longitudinal, MacArthur large strain
(From Tanner, 1962, p. 228, where sources of data are detailed)

The Human Growth Curve as a Primate Characteristic

The characteristic shape of the human growth curve is shared only
by apes and monkeys, the other members of the order *Primates*. It
is apparently a distinctive primate characteristic, for neither rodents
nor cattle have curves resembling it.

In Figures 9 and 10 are shown the velocity curves for body
weight of mice and chimpanzees (weight has to be used, since so
few data on length exist). In the mouse, as in the rat, there is little
interval between weaning and puberty, and no visible adolescent
spurt because there is no period of low velocity between birth and
maturity. In terms of the maturation of their organs, mouse and rat
are born earlier in development than man. The peak velocity of the
mouse's weight curve occurs at a time that corresponds closely, by
this organ-maturation calendar, with birth in man, which is when
the first peak of man's weight velocity curve occurs. In the chimpanzee (Figure 10), on the other hand, the curve resembles entirely

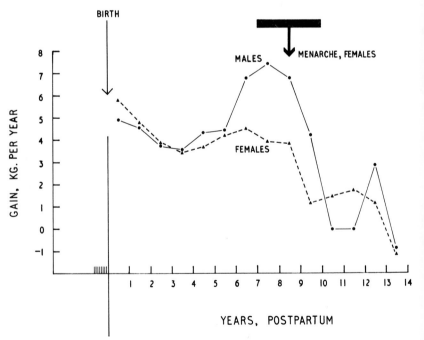

FIG. 10. Weight velocity curve for the chimpanzee. Actual increments, from Grether and Yerkes
(From Tanner, 1962, p. 234, where sources of data are detailed)

that of man. The first peak velocity of weight is shortly before, or at, birth, but this is followed by a gradual decrease of velocity during the long interval between weaning and puberty. The rhesus monkey has a similar curve, though with less time intervening between weaning and puberty.

The prolongation of the time between weaning and puberty appears to be an evolutionary step taken by the primates, reaching its most pronounced development in man. The increased time necessary for the maturing of the primate brain has been sandwiched into this period. At least some of the evolutionary reasons are not hard to find. It is probably advantageous for learning, especially learning co-operation in group and family social life, to take place while the individual remains relatively docile and before he comes into sexual competition with older individuals.

CHAPTER 2
Cells and the Growth of Tissues

All biological tissues are made up of cells. Even the skeleton is scarcely an exception, for the mineral deposits which are the characteristic of bone are laid down around an intricate network of bone cells, whose activity continuously adds and subtracts, remodels and repairs; if the bone cells die the skeleton soon becomes unserviceable. There are very many different sorts of cell in the body, but all have the same fundamental architecture and contain the same sort of machinery. Metaphors follow the economic and cultural activity of the age, and the modern picture of a cell is a factory with numerous production lines, each turned on or turned off by messengers arriving with instructions from the outside world.

Muscle and nerve cells fit this description only with some qualification. Most cells export their products, for use in particular bits of the general body economy; but muscle cells manufacture only their own form. Muscle is like a stack of playing-cards, with red and black cards stacked alternately. All the red cards project a little to one side, all the black cards a little to the other. Muscle movement corresponds to the cards sliding inwards again to make the edges of the pack even. It takes energy to pull the cards in in this way and it is this energy that is the product of the muscle-cell production lines. As for nerve cells, they do indeed have manufacturing lines producing substances which are extruded, but they also have the specialized function of electrical impulse transmission and this also requires energy.

Regenerating and Non-Regenerating Tissues
Muscle and nerve cells are peculiar also in that neither tissue is able to manufacture new cells once the growth period is over. When a cell is formed it lasts for most or all of the animal's lifetime. In many tissues, on the other hand, cells are continuously dying and being replaced by new cells. Examples are skin cells, blood cells and the cells lining the gut. Most such tissues have a special germinative zone where new cells are continuously manufactured.

In a third class of tissue, cells are relatively long-lived and stable but new cells can be made when the tissue is damaged, or when the work load on the tissue is greatly increased. Most glands (including the endocrine glands described in Chapter 7), and parts of the liver and kidney come into this category. In these tissues there is no specialized germinative zone; cells throughout the whole tissue, it seems, are capable of dividing when necessary, at any time throughout the animal's life. Thus liver has a marked power of regeneration; in the rat, a large proportion of it can be removed every month for a year or more, and every month the liver will regenerate itself. (Evidently the Greeks knew this, for when they created the myth of eternal punishment for Prometheus, it was his liver which the vulture pecked out each night. His pancreas would have done as well, but only that portion concerned with digestion and not the insulin-providing islets of Langerhans (p. 99); if the vulture had devoured them Prometheus would have died of diabetes.)

Muscle, however, has very little such power of regeneration, and nerve cells no power of regeneration at all. Nerve cells wear out (to use a vague term, indicating, perhaps, the occurrence of something analogous to metal fatigue), die and are lost. But they are not replaced.

The Dynamic State of the Body Constituents

All cells, however, have to renew their constituents, including even the constituents that make up the structural skeleton of the cell. There is a continual flow of substance into and out of the human body so that most molecules entering in the food stay only a relatively short time before being excreted. Nitrogen, for example, is mostly present in protein molecules, and an average adult man has a protein turnover of around 230 grammes per day; that is to say, he loses this amount, and replaces it with an equivalent amount of newly made protein. (Only a proportion of this is newly *ingested*, the rest being built up again from the remnants of broken-down protein: on a high protein diet the newly ingested proportion is about a half, while on a low protein diet it falls to a third or less; the absolute rate of turnover is little affected by diet.) Most of this turnover occurs in muscle, so that some 2 to 3% of muscle protein is replaced every day. In children the turnover is greater, especially during infancy, when the rate is about three times that of the adult. Different substances turnover at different rates; while the average nitrogen molecule remains in the body only a week or two, a molecule of calcium, the major constituent of bone, remains in the body a matter of months. A few substances, for example the cholesterol laid down in growing nerve sheaths, seem never to leave the body at all, but this is exceptional.

Thus the body is in a constant state of flux; we do not consist of certain particular molecules, but of a pattern imposed on continuously changing molecules. This dynamic state enables us to adapt to a continually changing environment, which presents now an excess of one type of food, now an excess of another; which demands different levels of activity at different times, and which is apt to damage the organism. But we pay in terms of the energy we must take in to keep the turnover running. Protein turnover accounts for much of our basal energy expenditure, i.e. the energy we produce as heat when totally at rest. Enough food must be taken in to provide this energy, or the organism begins to break up.

Cell Structure

The modern picture of a cell is a product of the electron microscope, which has revealed structural details quite unknown before. Figure 11 shows a diagrammatic reconstruction. The cell is surrounded by a membrane. Attached to the membrane are large numbers of receptor molecules, each specific for a particular chemical substance. Thus when a molecule of growth hormone (see pp. 91–3), for instance, passes by a particular type of liver cell, it is seized by its receptor on the liver cell membrane. Its capture starts a series of events in the cell which results in the turning on of the production line for a hormone called somatomedin (p. 93) which is then secreted into the blood stream. Thus the nature of the cell membrane is exceedingly important in the specialized function of the cell.

Inside the cell the most prominent structure is the nucleus. With few exceptions (see muscle, p. 30) cells have only one nucleus. Within the nucleus lie the chromosomes, invisible most of the time, because they are very long thin strands except when a cell is dividing. When a cell divides the chromosomes coil up like springs into short, thick and identifiable strands (see Figure 20, p. 53). These strands split longitudinally at cell division in a tissue like the liver, so that each daughter cell has the same chromosomes as the mother cell. Since the chromosomes control the fundamental nature and working of the cell this ensures that like cell begets like (except at one particular cell division, occurring only in germ cells, that need not concern us here). The chromosomes carry the genes,

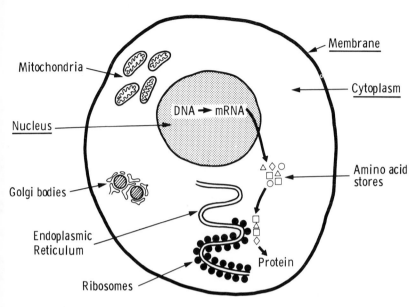

FIG. 11. Diagram of simple cell, to show important structures and assembly of protein molecules on ribosomes under DNA/RNA control

traditionally visualized as cylindrical beads strung, at irregular intervals, on the chromosomal threads.

Each gene consists of a length of DNA, deoxyribonucleic acid, a molecule resembling a rope ladder twisted into spiral form. The rungs of the ladder are of two different types. One is formed by a bonding of the substances thymidine and adenine, each substance sticking out from one side of the ladder. The other is similarly formed of two other substances, guanine and cytosine. Each rung may be arranged with one or the other end at the left- or right-hand side of the ladder; thus there are T–A, A–T, G–C and C–G rungs. When a strip of DNA multiplies the bondings break and the molecule, as it were, unzips. Half of the spiral ladder is free, with half-rungs hanging loose. A new piece is then manufactured exactly to correspond; only then can the molecule be zipped up again.

The rungs measure approximately one-millionth of a millimetre from one end to the other and are placed up the ladder at intervals of about a third of this distance. The ladder spirals gently, so it takes ten rungs for one complete turn.

A gene consists of a length of DNA containing typically about a thousand rungs. Since the rungs may be arranged in any sequence, the number of possible arrangements of a thousand rungs is practically infinite, and each gene has its own unique structure. It is the precise order of the rungs which constitutes the fundamental genetic instruction. Mutation is a change from one sort of rung to another. In many cases the number of mutations necessary to bring about the differences in structure between for example, bovine and human growth hormone (see Chapter 8) can be estimated fairly reliably. The affinities between different species and the speed of evolution are beginning to be thought of in these molecular terms.

It is important to realize that what is inherited is DNA, not blue eyes nor tallness nor even blood group A. The DNA of the chromosome acts as a template for the building of a similar molecule, RNA, ribosenucleic acid. This leaves the nucleus and enters the machine-shop of the cytoplasm (the rest of the cell outside the nucleus, see Figure 11). Here it activates that particular production line which assembles proteins from the amino-acids stored in some cellular pigeon-hole. Proteins consist of strings of amino-acids, often folded into extraordinary and beautiful shapes. There are only twenty amino-acids and millions of proteins, so the identity of a protein depends largely on the order and the number of its constituent amino-acids. It is this that the RNA controls. Each group of three consecutive rungs of the RNA ladder corresponds to, or 'specifies', a particular amino-acid (apart from groups of rungs which carry 'start-making' or 'stop-making' instructions). For this reason the system is often referred to as the 'triplet code' and each three-rung group is called a 'codon'. In this way the order in which the amino-acids are joined one to another, and hence the nature of the protein produced, is dictated by the RNA controller, and thus reflects the original instructions of the DNA. Thus a one-to-one relation exists, barring accidents, between the gene and the amino-acid sequence constituting the protein. Normally, therefore, we can say that blood group antigen A, which is such a protein, is directly gene-controlled, that is, inherited. Even in such cases, however, the working protein is often not quite the simple protein run off under the RNA controller, but a more complicated molecule, with additional pieces added to it by other sorts of cell machinery. Thus interactions between the gene product and the environment begin at the level of the cell machinery itself, and continue through the level of cell interacting with cell to the level of whole organism interacting with the outside environment.

The actual production lines appear to be the endoplasmic reticulum, a membrane-like structure along which are strung the ribosomes, analogous to work bays, at each of which a specific task of assembly is carried out.

Cell Growth and Organ Growth

An organ may enlarge either because the number of its constituent cells increases, or because the content of each of its cells increases, or because the amount of inert intercellular substance increases. In all non-regenerating tissues, growth goes through three phases (Figure 12). First, cell division and increase in cell numbers occurs, without any change in the amount of material in each cell. At this stage the tissue consists almost entirely of cell nuclei, with much intercellular fluid. The cells have little cytoplasm. In the second phase, the rate of cell division decreases; however, proteins continue to be synthesized at much the same rate as before. They now enter the cytoplasm and the cells become larger. The intercellular substance decreases correspondingly. New cells continue to be made, but at a slower rate than before. In the third phase, all division ceases and cells grow only in size. In all non-regenerating tissues, cell division (or mitosis) ceases some time before the cells stop increasing in size. The age at which cells cease dividing varies according to the tissue and organ considered.

Nerve. The first cells to stop dividing in the foetus are probably the neurons (or nerve cells proper) of the central nervous system, whose division is finished by about 18 postmenstrual weeks in the cerebral cortex (see Chapter 8, p. 106). The supporting tissue of the

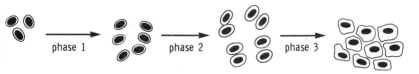

FIG. 12. Phases of cell and organ growth: phase 1, multiplication of all cells without addition of cytoplasm; phase 2, multiplication of some cells and addition of cytoplasm; phase 3, no multiplication of cells, further addition of cytoplasm

nerve cells, however – the neuroglia, which provide a scaffolding and serve the nerve cells in some such way, presumably, as bone cells serve bone – continue to multiply until one or two years after birth (see pp. 106–7).

Muscle. Muscles are exceptional in that muscle fibres, the basic units, are formed by fusion of several cells during development, a situation unique amongst developing tissues. In consequence, each fibre has more than one nucleus. The number of muscle fibres is fixed early in life, like the number of nerve cells. The size of the fibres increases, however, with the manufacture and incorporation of contractile protein. The number of nuclei seems also to increase, at least in man, keeping pace, more or less, with the increase in fibre size. At puberty in boys, both fibre size and number of nuclei increase considerably. However, these nuclei do not result from division of the existing muscle-fibre nuclei, which, once formed, like nerve-cell nuclei are incapable of dividing. Surrounding but outside the muscle fibres are numbers of nuclei which represent leftovers of the cells which formed the fibres in embryonic life. These cell nuclei are incorporated into the enlarging fibre, apparently differentiate into mature muscle-fibre nuclei, and then lose the ability to divide or differentiate further (see Chapter 10). Thus muscle carries a reserve supply of cells which even in adult life can provide a small potential for regeneration and repair. In this it resembles adipose rather than nerve tissue. It is not known for certain whether these reserve (or 'satellite') cells can multiply their number in childhood or in adult life in response to growth demand or injury, as can the cells of the cambium layer of the periosteum which covers bones (see p. 35). It is also rather uncertain whether all the nuclei newly appearing in muscle at adolescence are really in the muscle fibres themselves; their presence has only been inferred from chemical analysis, which shows an increase of DNA. DNA is present solely in nuclei; and all nuclei have approximately the same amount. Thus the amount of DNA tells one how many nuclei are present in a specimen. But it does not distinguish the nucleus of one type of cell from the nucleus of another, and some of the increase may represent connective tissue cells. The chemical analysis has to be viewed with caution; though its use enables many more experiments to be done in a given time, control of the results by electron and optical microscopy is still all-important.

Fat. Fat cells can be counted in biopsies of the buttock fat, obtained very simply by inserting a small needle and sucking fat up into it by means of an ordinary syringe. The fat cell is peculiar in that its nucleus is quite small, while its cytoplasm may enlarge to enormous bulk, carrying just a droplet of stored fat. The average size of fat cells does not appear to increase during pre-pubertal childhood, but the number of fat cells that can be counted increases steadily up to about the time of puberty. However, most present-day techniques for counting the number of cells depend on the cell having reached a certain size, i.e. being filled with a certain amount of fat. Thus it may be that fat is like muscle with embryonic nuclei left over in the tissue which cannot at present be identified or counted. As growth progresses, an increasing proportion of these may fill up with fat and become identifiable. If so, adipose tissue would be similar to muscle and nerve in that the period during which its nuclei could divide would terminate early in development, perhaps even before birth. This seems likely, since fat cells once formed do not appear to die, and in this they resemble muscle and nerve rather than regenerating tissues such as skin and glands.

Interference with the Cellular Clock

Professor Myron Winick, Dr Jo-Anne Brasel and their colleagues at Columbia University have shown in a long series of studies that severe undernutrition may stop cells dividing if it occurs at a time when cell division is normally going on. When the undernutrition is terminated cell division may proceed, but only at the rate appropriate to the age of the animal. This rate may be zero if the undernutrition has gone on long enough. Thus undernutrition may cause a permanent deficit in the number of cells. On the other hand, undernutrition occurring during the phase when cells are only increasing in size seldom has permanent effects. It seems that in this phase animals can simply stop the mechanism of protein storage in cells and then switch it on again, usually at a faster tempo than normal, when circumstances improve. We do not understand why the mechanism of cell division cannot be frozen in the same way (or at any rate to the same degree). The clock controlling it seems inexorably to continue whether or not materials are available to carry out the programme.

Whether overnutrition can cause an increase in the number of fat cells is not known. It has been said that overnutrition applied at the time when the cells are still being formed may indeed do so. However, as we have seen, it is not clear whether new fat cells are continuously formed during childhood, or whether existing embryonic cells are simply being filled up with fat to a level at which they can be seen. The issue is important, since it might be that the overfeeding of babies, prevalent in some parts of the world, induces an excessive number of fat cells as well as an excessive amount of fat in the cells. The distinction is crucial; hard though it may be to reduce the fat in an existing cell, it is at least possible, while actually to lose a fat cell probably is not. Thus obesity due to an excessive number of cells would be more intractable than obesity due to excessive amounts of fat in a normal number of cells.

Increase of muscle bulk may be regarded with more favour than increase in fat. Muscular training by heavy exercise does not cause an increase in the number of muscle fibres, but only in their size. However, training is applied (in man) at ages when cell division is believed to have ceased. It is an open question whether hormones have an influence on the number of muscle fibres in the foetus or even in the immediately postnatal child. If they do, it would be the androgenic or male-determining hormones such as testosterone which would be concerned (see Chapter 7), and it is possible that the effect might be permanent. Little research has been done on this aspect of athletics.

Growth of Bones
The manner in which bones grow merits a slightly longer description, both because it is bone lengths that we so often measure in our studies of growth, and because bone is an exceptional organ from the point of view of cellular growth.

A limb bone, such as the radius in the forearm, is laid down first as a cartilage model. At the centre of this model the cartilage cells break down and bone appears. This area is called the primary centre of ossification; in most bones it appears quite early in foetal life. Later, starting shortly before birth, secondary centres of ossification appear, mostly at the two ends of the long bones. They also represent conversions of pre-existing cartilage into bone, and are called epiphyses. New epiphyses appear in various parts of the skeleton right up to puberty.

The situation is then as shown in the diagram of Figure 13 and the first radiograph of Figure 14. The latter represents the distal or wrist end of the radius together with its epiphysis. The cartilage between these two bone areas is translucent to the x-rays and hence looks black in the radiograph. This area is called the growth-plate.

Cartilage cells in the part of the growth-plate immediately under

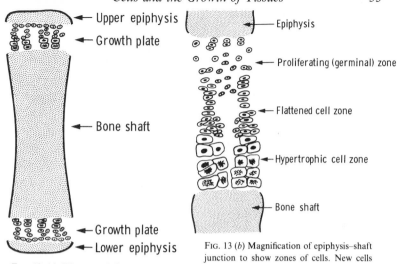

Fig. 13 (*a*) Diagram of limb bone with upper and lower epiphyses

Fig. 13 (*b*) Magnification of epiphysis–shaft junction to show zones of cells. New cells are formed in proliferation zone and pass to the hypertrophic zone to add to the bone accumulating on top of the bone shaft.

the epiphysis (the proliferative zone) divide and the cells so manufactured pass downwards into the growth-plate in longitudinally arranged columns of flattened cells looking like stacks of dinner plates (Figure 13b). As they pass towards the end of the main shaft of the bone, intercellular substance appears around them, so that each column is surrounded by a sleeve or tube. The cells enlarge and lose their flattened, stacked arrangement, and farther down still either they die, releasing mineral salts to the tube surrounding them, or they turn into bone cells (the matter is disputed). In this way the cartilage is replaced by bone. Thus a limb bone grows 'inwards' from its epiphysial end. In a radiograph of a rapidly growing bone (Figure 14a) the zone immediately next to the main shaft can be clearly seen, as a specially dense (i.e. white) band, presumably because it is there that calcium, opaque to the x-rays, is being laid down at maximal concentration.

As growth slows down the growth-plate gets thinner (Figure 14b). Finally bone from the main shaft breaks through it, reaches the bone of the epiphysis, and soon eliminates the plate altogether (Figure 14c). When the plate has gone, no more growth in limb bones is possible. The epiphyses are said to be closed, or 'fused' (i.e. joined to the main shaft).

In some animals such as rats the limb bone growth-plates may

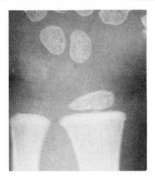

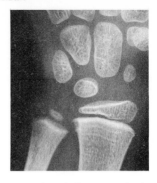

Fig. 14 (*a*) Radiograph showing epiphysis of radius at wrist; black area represents cartilage growth plate

Fig. 14 (*b*) Growth plate narrowed and zone of newly forming bone at end of shaft prominent as dense white band

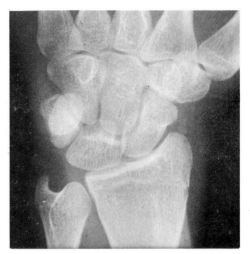

Fig. 14 (*c*) Growth plate practically eliminated, 'epiphyses fused with shaft'

never completely disappear, and growth may continue very slowly even into old age. In man it is the sex hormones which bring about the final elimination of the plates. Children who lack sex hormones remain with open epiphyses (and hence a pre-pubertal bone age – see Chapter 7). However, the capacity for such children to grow is not very great; growth is not stopped mechanically by closure of the epiphyses as used to be implied in older medical textbooks. All the sex hormones do is solidify an already quiescent area. Growth stops because the cartilage cells of the proliferative zone of the growth-plate gradually respond less and less to somatomedin, the hormone which causes them to enlarge and multiply. Why they do this is not clear. It may be that the quantity or affinity of the receptor simply decreases (see Chapter 7) or there may be a more active process of repression of the receptors by some other substance.

Most limb bones have epiphyses at both ends. However, the femur grows mainly through the activity of the epiphysis at the knee, and little by activity at the hip. The humerus, in contrast, grows chiefly at the shoulder end. The forearm bones grow chiefly at the wrist, and the tibia and fibula about equally at knee and ankle. The vertebrae of the spine have growth-plates like those in limb bones, one on the top and one on the bottom. The small round bones of the wrist (carpals) and of the foot have growth-plates too, but they are concentric and surround the bone; thus these bones grow continuously from outside in towards their centre.

The growth in width of limb bones takes place in a different way, without the intervention of cartilage. The bone is covered by a sheet of fibrous material called the periosteum, which consists of two layers. The outer layer is truly fibrous, but the inner, known as the cambium by analogy with tree growth (where it is the layer between bark and wood) consists simply of cells which multiply to lay down new bone; in this respect they are like the cartilage cells in the proliferative layer of the growth-plate. Thus the bone grows in width by new layers of bone being deposited on the outside of the existing bone. Growth of the bones of the skull and of most of the bones of the face also takes place in this way.

A growing bone has to be continuously remodelled; the intricate shape of an adult femur cannot be obtained merely by adding length and breadth to the infant femur. Large areas of bone have to be continuously broken down, reabsorbed into the body and lost. The main features of a bone's shape are controlled by genetic instruction, for a tiny cartilage precursor of a limb bone can be grown in an artificial medium or under the skin of the back (in experiments using the mouse) and the bone will grow and ossify into a respectable imitation of its appearance when in its proper place. The final modelling, however, seems to depend on muscle pulls and stresses from the remainder of the skeleton. Certain bones are more plastic than others in development, and respond to stresses and strains of near-by muscles with considerable changes. This applies in particular to the alveolar margins of the upper and lower jaws, i.e. the bones directly under the gums, which bear the teeth. It is absolutely essential for the animal to have teeth which fit accurately together and jaws which are adapted one to the other. In the wild, failure in this respect means starvation. It seems that such accurate fitting was better achieved by plasticity and moulding than by rigorous genetic control of both sides of the bite. Thus nature plays into the hands of the orthodontists who can move teeth about and even alter somewhat the upper and lower jaw bones by applying pressures, in a way mostly denied to the childhood orthopaedic surgeon faced with the task of unbending deformed limbs.

CHAPTER 3
Growth before Birth

The period of prenatal growth is vitally important to the child's future well-being; yet it is the period about which, inevitably, we know least. For the first and second thirds of pregnancy we have to rely on cross-sectional studies. In the second third we also have to rely almost entirely on foetuses expelled from the uterus because one or the other was abnormal, whereas during the earliest weeks of pregnancy we have the mostly normal products of social abortions. For later foetal life we can study infants born prematurely, making the assumption that these children have grown before birth and will grow after it in exactly the same way as children who remain in the uterus the average length of time. Though this may seem a hazardous assumption, it is probably justified, if children with certain abnormalities are excluded.

Our ignorance begins at the beginning, for we do not know what forces are responsible for selecting out of millions the one sperm which fertilizes the ovum. Fertilization takes place in one of the tubes which lead from the ovaries to the uterus. The fertilized egg spends some four to five days drifting down the tube and floating in the uterine cavity before it implants into the wall of the uterus. During this time the cells divide steadily, so that by the time of implantation the blastocyst, as it is called, consists of around 150 cells. Blastocysts can be washed out of the tube and implanted in foster-mothers, and in animals this technique has been used both for examining the effect of different kinds of uterine environment on development and for transporting by air the foetuses of large animals packed, temporarily, in the uteri of small ones.

After implantation the outer layer of the blastocyst undergoes a series of changes which culminate in the formation of the placenta. A small proportion of the inner layer develops into the embryo. The period of the embryo is considered to begin 2 weeks after fertilization and ends 8 weeks after fertilization, when the child, now recognizably human, with arms and legs, a heart that beats, and a nervous system that shows the beginnings of reflex responses to tactile stimuli, is called a foetus. At this stage it is about 3 cm. long (see Figure 15).

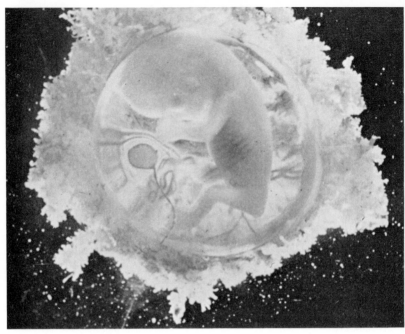

Fig. 15. Foetus at 13 postmenstrual weeks
(From D. W. Smith, 1977, by kind permission)

This is a hazardous period, and many more ova are fertilized than come to fruition. Some 10% fail to implant and of those which implant and become embryos about 50% are spontaneously aborted, usually without the mother knowing anything has happened. Such abortion is due in most cases to abnormalities of development, either of the embryo or of its protective and nutritive surrounding structures. Abnormalities of the chromosomes, for example (see p. 209) are present in 5% to 10% of fertilized ova but in only 0·5% of newborns. Thus 90% to 95% of all conceptuses with these abnormalities are rejected, as early spontaneous abortions. Lesser degrees of abnormality may result in a viable foetus and child, but one which grows less than is normal. An example of this is the child with the disorder called Down's syndrome, caused by possession of an abnormal number of chromosomes (see p. 209).

The reckoning of age in the prenatal period presents a problem. Traditionally, and because we have no better way, age tends to be counted from the first day of the last menstrual period. This occurs on average 2 weeks before fertilization. Thus the most frequent age at birth is 280 days or 40·0 weeks, reckoned as 'postmenstrual age', but this represents only 38 weeks of true foetal age. There are difficulties in individual cases, however; the interval from menstruation to fertilization varies considerably; and worse, menstrual bleeding, or something like it, may continue in some women for 1 or even 2 months after fertilization. In studies of organ differentiation and development, postfertilization age is frequently used (see, for example, the next chapter).

Reliable growth curves of the foetus are hard to come by, for the reasons given at the beginning of this chapter. There are good modern data for body length of foetuses from about 10 to 18 weeks postmenstrual age, and for infants born prematurely from about 28 weeks onwards. In between 18 and 28 weeks there are almost no useful data. Figure 16 is therefore somewhat diagrammatic; it shows the distance and velocity curves of body length, so far as they may be measured in prenatal life, and for the first year after birth. The position of peak velocity is somewhat uncertain. The solid lines represent the actual length and length velocity; the interrupted lines the theoretical curve which might be supposed to occur if no uterine restriction took place in the last weeks of pregnancy, followed by a compensating catch-up after birth (see p. 42).

In the embryonic period the velocity is not very great. During these first 2 months differentiation of the originally homogeneous whole into regions, such as head, arms and so forth occurs, as does histogenesis, the differentiation of cells into specialized tissues such as muscle and nerve. At the same time each region is moulded into a definite shape, by differential growth of cells or by cell migration. This process, known as morphogenesis, continues right up to adulthood and, indeed, in some parts of the body, into old age. But the major part of it is completed by the 8th postmenstrual week.

The high rate of growth of the foetus compared with that of the child is largely due to the fact that cells are still multiplying. The proportion of cells undergoing division in any tissue becomes progressively less as the foetus gets older, and it is generally thought that few, if any, new nerve cells (as distinct from the supporting neuroglia) and only a small proportion of new muscle cells appear after 30 postmenstrual weeks, by which time the velocity in linear dimensions is dropping sharply. The muscle and nerve cells of the foetus, however, differ considerably in appearance from those of the child or adult, for early in development there is little cytoplasm around the nuclei (see Figure 12, p. 29). In foetal muscles there is a

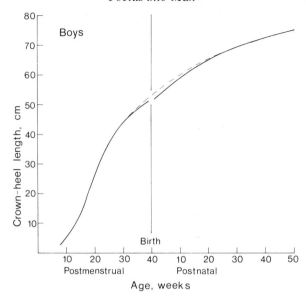

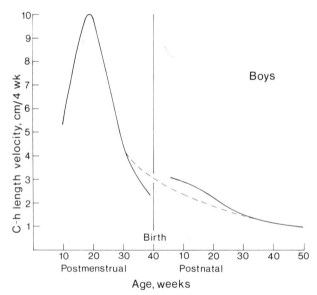

Fig. 16. Distance (*above*) and velocity (*below*) curves for growth in body length in prenatal and early postnatal period. Diagrammatic, based on several sources of data. The solid lines represent actual length and length velocity; the interrupted line the theoretical curve if no uterine restriction took place.

(Tanner, unpublished)

great deal of intercellular substance and a much higher proportion of water than in mature muscle. The later foetal and postnatal growth of muscle consists chiefly of building up the cytoplasm of the muscle cells; salts are incorporated and the contractile proteins are formed. The cells become bigger, the intercellular substance largely disappears and the concentration of water decreases. This process continues quite actively up to about 3 years of age and slowly thereafter; at adolescence it briefly speeds up again, particularly in boys, under the influence of androgenic (male-determining) hormones. In the foetal nerve cells cytoplasm is added, and the cell processes grow. Thus postnatal growth, for most tissues, is chiefly a period of development and enlargement of existing cells, whereas early foetal life is a period of cell division and the addition of new cells.

During the last 10 weeks in utero the foetus stores very considerable amounts of energy in the form of fat. Up till about 26 weeks postmenstrual age most of the increase in foetal weight is due to accumulation of protein as the main cells of the body are built up. But from then on fat begins to accumulate, both deep in the body and subcutaneously. Drs D. A. T. Southgate and E. Hay (1976) have analysed some 36 foetuses of this age and found that from about 30 to 40 postmenstrual weeks fat increases from some 30 g to 430 g. Since fat packages much more energy than protein or carbohydrate per unit volume this represents a large reserve of energy available for the first, perhaps critical, period after birth (newborn monkeys do not store up fat in this way, hence look less cuddly to the human eye). Conversely, the creation of such a store represents a considerable drain on the energy resources of the mother in the last weeks of pregnancy.

The Effect of the Uterine Environment

Growth in weight of the foetus follows the same general pattern as growth in length, except that the peak velocity is reached later, at approximately 34 postmenstrual weeks. There is considerable evidence that, beginning at 34 to 36 weeks, the growth of the foetus slows down owing to the influence of the uterus, whose available space is by then becoming fully occupied. Twins slow down earlier, when their combined weight is approximately the 36-week weight of the singleton foetus. In Figure 17 is plotted a velocity curve constructed, before birth from the differences between weight means of singleton babies born alive at periods from 24 to 40

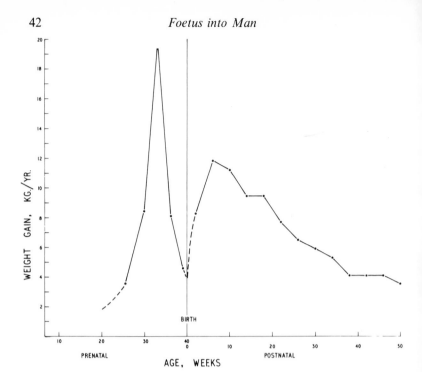

FIG. 17. Velocity of growth in weight of singleton children. Prenatal curve drawn from data of McKeown and Record (1952) on birthweights of live-born children delivered before 40 weeks of gestation. Postnatal data from Ministry of Health (1959); mixed longitudinal data (their Table VII). Dashed line shows estimate of velocity immediately before and after birth, showing catch-up.
(From Tanner, 1963)

weeks, and after birth from differences in weight means in a mixed longitudinal study of a similar population in the U.K. A curve comparable in shape with usual growth curves could be constructed by joining the top, 32-week prenatal point with the first, 8-week postnatal point. Such a hypothetical curve would predicate a peak weight velocity reached at about 34 weeks. But between then and 40 weeks, growth is held up; and the increase in velocity in the first 8 weeks after birth represents a catch-up (see Chapter 10), on the part of those newborns who have been most delayed in the uterus. Thus there is a significant negative correlation between length at birth and length gain in the first 6 months after birth, and also between weight at birth and weight gain in the first 6 months. The smaller the baby, on average the more it grows at this time.

This slowing-down mechanism enables a genetically large child developing in the uterus of a small mother to be successfully delivered. It operates in many species of animal; the most dramatic demonstration was made by crossing reciprocally a large Shire horse with a small Shetland pony. The pair in which the mother was the Shire had a large newborn foal, and the pair in which the mother was the Shetland had a small foal. But both foals were the same size after a few months, and both ended about half-way between their parents. The same has been shown to occur in cattle crosses.

A frequently used measure of the degree of association between two variables is called the *correlation coefficient*. It varies from 0·0, which means the two variables are quite independent, to 1·0, which means that one is totally predicted by the other. In man the correlation coefficient between length at birth and adult height is only about 0·3; but the coefficient rises sharply during the first year, and between length at age 2 and adult height is nearly 0·8. These figures also reflect maternal control of newborn size.

How this control is exercised is not clear. The placenta grows at first more rapidly than the foetus, but from about 30 weeks onwards its growth rate becomes less than that of the foetus and the placenta/foetus ratio falls. It may be that the placenta simply cannot increase its capacity to supply enough food to sustain the rapid 34-week foetal velocity. In mice and guinea pigs it seems likely that the limiting factor is blood flow, the size of the placenta depending on the pressure at which the maternal blood reaches it, and the size of the foetus depending, in turn, on the size of the placenta. Whether this is also important in man is not yet known.

Poor environmental circumstances, especially of nutrition, result in lowered birthweight in the human. This seems chiefly to be due to a reduced rate of growth in the last 2 to 4 weeks of foetal life, for mean weights of babies born at 36 weeks to 38 weeks in various parts of the world under various circumstances are rather similar. Mothers who, because of adverse circumstances in their own childhood, have not achieved their full growth potential may produce smaller foetuses than they would have done had they grown up under better conditions. Thus two generations or even more may be needed to undo the effect of poor environment on birthweight. In Guatemalan villages, for example, mothers of short stature had babies of lower birthweight than did mothers of medium stature. A food supplement given to the mothers during pregnancy had more effect on the birthweight of children born to short mothers than on those born to medium-sized mothers, but it did not wholly eliminate the difference.

So-called 'Premature' Babies

The average length of gestation, measured from the first day of the last menstrual period, is 280 days or 40 weeks. However, there is considerable individual variation around this figure (even when mistakes and inaccuracies in determining gestational age are set aside), and lengths of gestation from 259 days (37 completed weeks) to 293 days (42 completed weeks) are by international agreement regarded as normal. Babies born within these limits are called *term* babies. Babies born earlier are called *pre-term* babies and those born later *post-term* babies. Recent investigations lead to the conclusion that the limits of normal term should really have been set between 38 and 41 completed weeks instead of between 37 and 42 weeks.

Until a few years ago all babies who weighed less than 2,500 g ($5\frac{1}{2}$ lb.) at birth were designated 'premature' whatever their length of gestation or physiological state. This definition (promulgated by WHO in 1948) caused much confusion and has now been dropped; the word 'premature' has disappeared from scientific use. Babies less than 2,500 g at birth (WHO, 1961) are called 'low birthweight' babies; this low birthweight may be due either to their being born early, or to their being babies who are pathologically small for their length of gestation. The distinction is made by the use of standards such as the one shown in Figure 18, which gives the centiles for first-born girls only; children subsequent to the first are heavier (on average by 110 g) and boys are heavier than girls (on average by 150 g). The baby marked A weighed 2,500 g but was born a month early, at 36 weeks. She is at the 25th centile for first-born girls of this gestational age, thus perfectly normally grown. The baby B, also of 2,500 g, born at 40 weeks (full-term), is below the 5th centile and thus of very questionable normality. Such a baby is known colloquially as 'small-for-dates'.

Since there is a tendency for some mothers always to have relatively small babies and other mothers relatively large ones, a more critical standard can be constructed in which the weight of the new baby is compared with the birthweight of his brothers and sisters (length of gestation being allowed for). To a large extent this family trait of size at birth is inherited, probably through characteristics of the maternal uterus rather than of the foetus itself. This

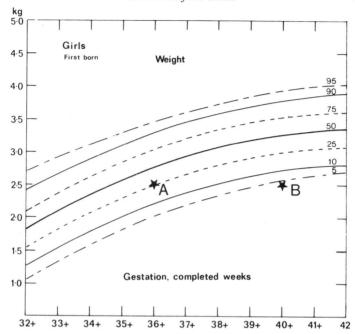

Fig. 18. Standards for birthweight for gestational age, first-born girls: (*a*) 2,500 g, born at 36 weeks; (*b*) 2,500 g, born at 40 weeks
(From Tanner and Thomson, 1970)

variation between families, however, remains strictly within the normal range of birthweights. When we talk about small-for-dates babies we mean those who are beyond these limits. A small overlap exists, for a very few normal babies are necessarily below the arbitrary limit set for the upper bound of small-for-dates infants. This occurs in the use of all statistical standards, as explained in the Introduction.

The distinction between pre-term babies of normal weight for length of gestation and babies who are light for their (often normal) length of gestation is an important one. The purely pre-term infants catch up perfectly well and seem little the worse for their earlier experience of the outside world. Even those born as early as 28 weeks weighing 1,000 g can nowadays be kept growing at the very rapid rate appropriate to their age and sent home 8 to 10 weeks later at the normal weight for a full-term infant. This follows the recent introduction of concentrated milk preparations together with intravenous nutritional supplements to provide the necessary amounts of calories and protein.

On the other hand, babies who are light for length of gestation (small-for-dates) do not on average catch up to normals, although they diminish the gap a little. The average small-for-dates child reaches about the 25th centile for height, which implies that a considerable proportion of such children remain below the 3rd centile limits of normal. A considerable proportion fail also to develop the same level of mental ability as normal children.

Children can of course be both pre-term and small-for-dates; about one-third of pre-term babies in the U.K. appear to be so. Such children fall clearly in the small-for-dates category. The deficits in later size and ability get worse as the birthweight decreases; children between 2,000 and 2,500 g at (full-term) birth show little impairment of ability and only a slight size deficit. A considerable proportion of those under 2,000 g, however, have some mental or neurological defect. These figures concern, of course, survivors. The chance of perinatal death (i.e. death within 24 hours after birth) increases, the lower the weight for gestation. In a study of 44,000 consecutive births in the Royal Victoria Hospital in Montreal, Usher and McLean found that the perinatal mortality was 54 per 1,000 births in babies born at term (37 to 42 weeks) with weights under the 3rd centile for gestational age compared with 6 per 1,000 in babies with weights between the 30th and 70th centiles. In babies of all lengths of gestation, perinatal mortality was 16 per 1,000 amongst babies whose weights were between the 3rd and 97th centiles for gestational age, but 189 per 1,000 for those whose weights were under the 3rd centile for gestational age.

Thus the prognosis for a small child born after a normal-length gestation is very different from the prognosis for an equally small child born after a shortened gestation. Leaving the uterus early is not in itself harmful, whereas growing less than normally during a full uterine stay implies pathology of foetus, placenta or mother.

Small-for-dates babies are a heterogeneous group, produced by several different causes. In countries where severe maternal malnutrition occurs a proportion of small-for-dates babies is due to this. But the malnutrition has to be fairly severe, for in this situation the foetus is protected at the expense of the mother. In well-nourished mothers disorders of the placenta may be responsible. Smoking in pregnancy causes, on average, a reduction of 180 g in full-term foetal weight and a 30% increase in perinatal mortality; the size of reduction persists throughout childhood. This reduction of weight brings some babies into the small-for-dates category. Smoking presumably works by affecting placental blood flow and foetal nutrition, though the less likely possibility of direct action on the foetal cells cannot be ruled out. Alcohol appears also to reduce foetal weight, and a large intake of alcohol may affect the foetus directly, causing a recognizable disorder known as the foetal alcohol syndrome, described by Professor David Smith and his colleagues of the University of Washington, Seattle. The face of the baby has a characteristic appearance, due to insufficient development of parts around the eyes, nose and upper lip. Some maternal diseases cause smallness-for-dates; in particular rubella (German measles), which also may cause specific deformities and deafness. In other instances it seems the foetus itself is disordered and lacks the capacity to grow properly. According to some studies, mothers of small-for-dates babies have a higher proportion of abnormal outcomes in other conceptions (miscarriages, etc.) than do mothers of normal-sized babies.

Since the worst period for the small-for-dates baby seems to be the last part of pregnancy, when he is, or should be, growing most, many doctors are now advocating removal of these foetuses at 36 weeks, or even at 34 weeks in severe cases. This is done by straightforward induction of labour and vaginal delivery. Much research is therefore in progress to devise means of recognizing these foetuses in time for this action to be taken. To date, the most helpful guide to intra-uterine growth insufficiency is measurement of the foetus by ultrasonic means. Ultrasound may be used early in pregnancy to measure the length of the foetus's back from crown of head to rump, and from about 13 postmenstrual weeks to measure the width of the head and the circumference of the abdomen. It seems that screening all babies for crown to rump length between 6 and 12 weeks, for head width between 13 and 20 weeks, and for abdominal circumference at 32 weeks, would detect over 90% of all small-for-dates babies, with a negligible number of false positives (i.e. babies so diagnosed who were really normal). Such a screening programme is well within the capabilities of the medical services of industrialized countries. Biochemical tests for the integrity of placental function are being developed using the mother's blood or urine, but at the time of writing are not yet ready for general use.

Head Growth in Utero

Measurements of foetal weights have necessarily to be cross-sectional, but the advent of ultrasonic scanning has made possible fully longitudinal studies of the growth of the foetal head. Very-high-frequency sound waves (too high to be heard) are beamed at the foetus and echoed back from the junctions of differently conducting tissues. Thus the waves bounce back from the hard foetal head and the time taken for the echo to return indicates the distance from source to the edge of the skull. By scanning with a number of beams, and processing the results by computer, the foetal head may be outlined and accurate measurements of its maximal transverse diameter obtained without any harm being done. Foetal length from head to rump, and foetal abdominal circumference, can often be estimated also, though with less accuracy.

Figure 19 shows two individual curves of head width from 20 to 40 weeks, kindly made available by Professor Stuart Campbell of King's College Hospital, London University, who played a leading part in developing the ultrasonic method. The two curves are plotted against standards of head width of normal foetuses, and both come from children of low birthweight. Curve A shows a foetus with normal growth until 36 weeks; curve B a foetus whose head growth was abnormal from a very early age. Professor Campbell, with his paediatric colleagues, has followed up some 60 small-for-dates children to an average of 4 years. All these had had serial measurements of head growth from 30 postmenstrual weeks or before. The results showed unequivocally that those small-for-dates children whose head growth was normal, or whose velocity of head growth had only started to fall after 34 postmenstrual weeks (Figure 19a), were normal in body size (or very close to it) and in intellectual development at follow-up. Those whose head-growth failure started before 34 weeks, and thereafter continued, were much smaller at follow-up, and had lower-than-normal intellectual attainment. Studies such as these are beginning to provide new insight into foetal growth and its pathology.

Birth as a Happening

For some physiological functions birth signifies upheaval and change, often associated with a particular vulnerability. But it is important to realize that for very many others birth is an incident without much significance in a steadily changing and maturing programme of events.

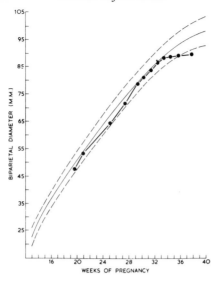

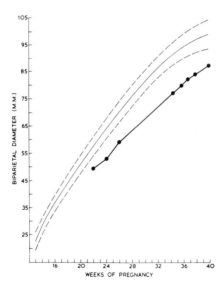

FIG. 19. Growth of head width in two boys, measured by ultrasound scan. Chart shows mean and limits (3rd and 97th centiles) for each completed week of gestation: (*above*) normal growth till 34 weeks, then marked slowing; labour induced at 38 weeks, birthweight 2,450 g; (*below*) head already small at first measurement (22 weeks); slow growth began at 18 weeks or before. Birthweight, 1,890 g (From Campbell (1976) and Fancourt *et al.* (1976), with permission of authors and *British Medical Journal*)

The respiratory and cardiovascular systems are the ones most altered by the fact of birth. Failure to establish satisfactory respiration during the crucial period just after birth is a common cause of neonatal death and of brain injury amongst survivors. The newborn infant, however, has a greater capacity to survive straightforward lack of oxygen without damage than have children or adults, and neonatologists are moving towards the view that in many cases it is pre-existing brain damage, of developmental or uterine origin, which is the reason for failure to start breathing. 'Almost all medical literature still equates failure to breathe at birth with "brain injury", placing the damage at the time of delivery. The evidence suggests that pre-existing brain damage interferes with the ability to adjust to extra-uterine life and to initiate respiration', wrote Drs H. Knobloch and B. Pasamanick (1962) in a review of mental subnormality. Dr C. M. Drillien (1964), whose study throughout childhood of babies in Scotland born small-for-dates is amongst the most thorough, remarks that the most severe defects originate usually at an early stage of foetal development rather than from damage to a potentially normal nervous system occurring in the last 3 months of pregnancy or during delivery. However, differences in later behaviour were found between children who had been subjected to certain obstetric hazards and those who had not, and these might have been wholly or partly due to minimal brain damage occurring at the time of delivery. The success or otherwise of the early delivery of small-for-dates babies (see p. 46) will show which view is predominantly correct.

Some enzyme systems are sharply affected by the fact of birth. Thus in rabbits there is a sudden rise in glucose-6-phosphatase in the liver at birth and not before, whether birth is induced early, occurs normally, or is late. This is readily accounted for. The enzyme catalyses the breakdown of glycogen to glucose in the liver. Just after birth there is a great demand for glucose in the blood and tissues, pending the full establishment of lactation, and the enzymic events are adjusted in a way which satisfies this demand.

On the other hand, numerous enzymes exist with time courses of development which are independent of birth. A well-known example concerns the switch from production of foetal haemoglobin to adult haemoglobin. These are forms of haemoglobin with slightly different molecular structures. A gradual switch in production occurs, so that the percentage of foetal haemoglobin begins dropping from about the 36th postmenstrual week. The switch is not due to birth, however, because the percentage of foetal haemo-globin remains high in pre-term babies and is low in those who are born post-term. Thus in this, as in many other respects, the prematurely born has to await the striking of some previously regulated biological clock.

Most important, the maturation of the nervous system seems to be little affected by the fact of birth. The electroencephalogram (or 'brain wave' record) of the infant born at 28 weeks will 6 weeks later be much the same as that of an infant born at 34 weeks, provided adequate care is given. Dr C. Dreyfuss-Brisac, the fore-most expert in this field, wrote (1975): 'It is particularly important to point out that electroencephalographic maturation is closely and exclusively related to conceptional age ... the duration of extra-uterine life is not an important factor in EEG maturation.' Birth also fails to influence the date of appearance of conditioned reflexes or the stages of motor behaviour elicited by detailed neurological examination. According to Dr St Anne Dargassies (1966) the development of motor behaviour occurs in a pre-term infant 'just as if birth before term did not perceptibly alter the course of neuro-logical maturation. [The schedule] is adhered to just as closely in an incubator as in utero.' A strikingly simple example is given by the rates of conduction of nerve impulses along the peripheral nerves in the arm. The rate gets gradually faster up to about 2 years of age, when adult values are reached. The writer's colleagues Professor Dubowitz and Dr Moosa of the Institute of Child Health found that at 40 weeks after conception only very slight differences in conduction rate existed between normal-for-gestation pre-term babies born at 36 weeks and earlier, and normal babies born at 40 weeks. Even this slight difference had disappeared by 45 weeks postmenstrual age.

CHAPTER 4
Sex Differentiation up to Puberty

Chromosomal Sex

It is possible to take human white cells from the blood, or connective tissue cells from a tiny piece of skin, grow them outside the body and stop their activity at the moment in their division when the chromosomes are especially visible. The chromosomes, which are in the nucleus of the cell, are then stained and the cell is photographed. The chromosomes are the structures which carry the genetic blueprint, in DNA molecules (see p. 27) which specify whether the cell belongs to a man or a mouse, to a blue-eyed or brown-eyed man, a man who under favourable childhood circumstances will grow to be 6 ft tall or a man who under the same circumstances will grow to be $5\frac{1}{2}$ feet.

The chromosomes also specify whether the cell belongs to a male or a female. Chromosomes come in pairs; in a given cell of a child one chromosome of each pair is derived from the father and one from the mother. There are 23 pairs of chromosomes in man, and the members of each pair are distinguishable from members of all other pairs by their size, shape and staining patterns. When the chromosomes are stained in the cell nucleus they lie higgledy-piggledy and not paired up with their opposite numbers, so for clarity it is customary to take photographs, cut out each individual chromosome and then arrange them in pairs as shown in Figure 20. The chromosomal pattern thus displayed is called a *karyotype*. The members of each pair of chromosomes appear identical and indeed carry precisely corresponding pairs of genes, though in many of the pairs the two genes are in slightly different forms. There is one major exception, however; the members of one pair of chromosomes are grossly dissimilar. One, the x chromosome, is amongst the largest of the whole set, and the other, the y chromosome, is amongst the smallest. These are the chromosomes that specify sex, for in all mammals creatures with two x chromosomes are females and those with one x and one y are male.

This is where sex differences begin. Ova and sperm are special cells, and contain only one member of each chromosome pair. Thus when they join at fertilization the fertilized egg has the usual pairs

XY

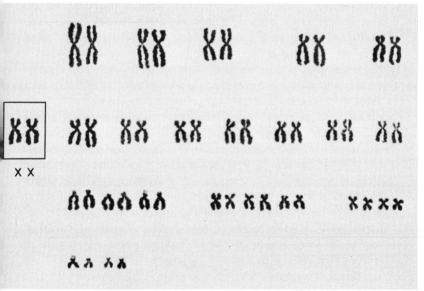

X X

Fig. 20. Karyotypes of normal male (*above*) and female (*below*)
(Reproduced by kind permission of Professor Paul Polani)

of chromosomes, one member of each derived from each parent. Since the karyotype of the father is x y a sperm may carry either the x or the y. The karyotype of the mother is x x so the ovum carries one or the other of these xs. The sex of the child therefore depends on the sex of the sperm. When the fertilized egg divides, it passes a pair of chromosomes on to both daughter cells, which are thus x x from a female fertilized ovum or x y from a male fertilized ovum. The same occurs at all subsequent divisions (except those making new ova or sperm), so all the cells of the child have the karyotype x x if a girl, x y if a boy.

Sometimes this process goes wrong, usually because a pair of chromosomes fails to split cleanly into the singles needed for ovum and sperm. Mistakes may happen with any of the chromosomes but seem to be particularly common in the sex chromosomes. In as many as 1% to 2% of all conceptions a fertilized ovum is made with a single x chromosome only, without either a y or a second x. The karyotype is written x o. The very great majority of such children die in utero, often so early in pregnancy that the mother does not even know she has aborted an embryo. The few that survive (including mosaics, see below) constitute about one out of every 3,000 female babies. They are girls in external appearance, but their ovaries degenerate early in childhood so they cannot produce the female sex hormone at puberty which causes the breasts to grow. They are, of course, infertile. They are also strikingly short in stature, owing to a diminished response of the bones to the hormones controlling growth. The appearance of their face, neck, shoulders and chest is characteristic, though sometimes only distinguishable by the expert. The condition is known as gonadal dysgenesis or Turner's syndrome. Sometimes the defect in the mechanism of chromosome duplication occurs at a later stage than initial fertilization. Then there may result a child with two lines of cells, some being x o descendants of the erroneous division, and some normal x x. Such children are called *mosaics*; in appearance and function they may be anywhere between the full Turner's syndrome and normal girls. A much higher percentage of these mosaic individuals survive the foetal period and they constitute about a quarter of the surviving children with Turner's syndrome.

Other sex chromosome errors also occur. Children with the karyotype x x y (Klinefelter's syndrome) occur about once in every 1,000 male live births; they are boys with characteristically long arms and legs, small testes, reduced fertility and often a slight degree of mental handicap. Other chromosomal mistakes produce such karyotypes as xxx, xxxy and xyy. The last anomaly, which also has an incidence of about one per 1,000 male live births, produces particularly tall men. Boys with xxy and xyy karotypes nearly all survive foetal life.

A similar mistake in chromosome No. 21 (see Figure 20), giving three instead of two 21s, is responsible for the well-known form of physical and mental handicap called Down's syndrome or mongolism. Chromosome 21 is small and presumably for this reason many triple-21 children survive; triplication of other chromosomes nearly always causes foetal death. It seems that these chromosomal errors occur more frequently in older mothers than in younger ones. The recent lowering of the average age of child-bearing in the U.K. has dramatically reduced the incidence of Down's syndrome.

Prenatal Differentiation

During the earliest stages of development only examination of the chromosomes tells us whether the embryo is male or female. When the gonad (a term including both testes and ovaries) first appears, it looks the same in both sexes. Its rate of development, however, is greater in the male. This is presumably due to genes on the y chromosome, though we do not know the intermediary steps by which they exert their effect. Between 7 and 8 weeks after fertilization the male gonad becomes recognizable as a testis; not until 2 weeks later does the female gonad undergo the changes which identify it as an ovary. About 9 weeks after fertilization, Leydig cells appear in the testes (see Chapter 7) and begin to secrete testosterone. The testosterone causes the previously undifferentiated external genitalia to form a penis and scrotum. This formation is complete by about 13 weeks after fertilization; thereafter the penis grows faster than the clitoris to produce the difference seen at birth. The stimulation of the Leydig cells during the period of formation of the genitalia is due to a hormone coming from the cells of the placenta (chorionic gonadotrophin). During the subsequent period of penis growth, testosterone levels are kept elevated by the same hormone plus perhaps some stimulation from the foetus's own pituitary gland. The testes also secrete a second hormone, from the Sertoli cells, which causes atrophy of the structures that in females develop into the Fallopian tubes, which carry eggs from the ovaries to the uterus.

In the absence of testosterone stimulation the external genitalia develop at around 12 weeks after fertilization into those of the female, apparently quite passively, without specific hormonal induction. It seems to be generally true throughout sex differentiation that the female is the 'basic' sex into which embryos develop if not stimulated to do otherwise. Even in the last stages of differentiation, at puberty, the male changes are more striking and extensive than those of the female, although in this instance girls as well as boys do have their own hormonal stimulus.

These changes lead to a qualitative difference – indeed to the only qualitative difference – between the sexes. Barring accidents of development the internal and external genitalia are always either male or female. Other sex differences – of body size, shape and tissue composition; of rate of maturing, strength and numbers of red blood cells and so on – are quantitative. Though a large difference in height, for example, exists between the average man and woman, there is overlap of the distributions, so that tall women are taller than short men.

The beginnings of many of these quantitative differences are traceable to the prenatal period. For example, the length of the forearm relative to the upper arm or to total body height is greater in the average male than in the average female. This sex difference begins in utero and continues very gradually to develop throughout the whole period of growth. The same is true of the intriguing and, from an evolutionary point of view, rather mysterious difference between the relative lengths of index and fourth fingers. The index finger is longer than the fourth more frequently in females than in males, and this pattern can already be seen at birth. (To make matters still odder, it is necessary to add that the right side of the body is on average more male in this matter, and indeed some others, than the left side; thus a man usually has a relatively longer fourth finger on the right hand and a relatively longer index on the left hand, a woman the reverse.)

Brain Differentiation

From an evolutionary point of view it is little use differentiating a peripheral organ such as the testis or ovary unless the messages causing them to function are differentiated also. The messages to the ovary cause it to operate cyclicly (hence the menstrual cycle); those to the testis cause it to operate non-cyclicly. These messages are hormonal and come from the pituitary gland (see p. 90). But they are initiated and controlled by a portion of the brain called the hypothalamus which instructs the pituitary to send out its various hormones. Not only, then, has the endocrine system to be differentiated according to sex, but the hypothalamus also. It is known that in the rat, guinea pig, and many other species it is, as usual, the male brain which is changed, by testosterone secreted by the foetal testes. The change, involving cell structure, once made, is irreversible. One can transplant a pituitary gland from an adult female rat into a male whose own gland has been removed and the transplant, when it starts to work, will obey the brain and function as male. In other words, endocrine maleness resides in the hypothalamus and not in the testis or the pituitary.

Recent work has uncovered an amusing irony about this differentiation. Testosterone, the male sex hormone, has actually to be turned into oestradiol, the female sex hormone, in the cells of the hypothalamus, before it can exert its action (chemically speaking, it is not a very far cry from one hormone to the other). It is the oestradiol which causes the cells to grow connections that link together in a wiring pattern that is irreversibly male. Only testosterone-derived oestradiol can do this, because in the young female animal oestradiol circulating in the blood is kept out of the brain cells by being bound to a special protein in the blood. Endocrinology (discussed in Chapter 7) is full of such intricate surprises.

Furthermore, in the rat, brain differentiation is a prime example of a sensitive period (see pp. 161–3). The testosterone must be secreted during the first 5 days after birth for the brain to differentiate. After this, it is too late; before this, it is too early and nothing happens. The rat is born relatively much earlier than the human, so this period corresponds to 12 to 14 postfertilization weeks (or 14 to 16 postmenstrual weeks) in man. It is known that blood testosterone level in the male foetus is high at this time but it is all the same uncertain whether differentiation exactly similar to that in the rat occurs in man. Probably it does, but some recent evidence in monkeys suggests that primates (monkeys, apes and man) may be somewhat different in the way they regulate their reproductive affairs, even at this level.

In the rat, there is evidence that a second, higher, brain centre, controlling types of mating behaviour, is also irreversibly differentiated. The control seems to be more quantitative than qualitative, which implies that only a partial reversal of sex role can be brought about by experimentally changing the animal's endocrine apparatus. It is at present unclear whether this second centre is present in man. If some such differentiation of the brain does occur, occasional errors in the process would certainly be expected. They would probably be quite rare, however, and could scarcely account for more than a small fraction of human sexual pathology.

Differentiation in Childhood

From birth till puberty sex differentiation continues, in the physical no less than in the psychological sense. The most important difference is in the rate of maturing.

Girls grow up faster than boys: that is, they reach 50% of their adult height at an earlier age (on average at 1·75 years compared with 2 years in boys), enter puberty earlier and cease earlier to grow. The difference in tempo of growth starts very early; half-way through the foetal period the skeleton is already some 3 weeks more advanced in girls than in boys. At birth the difference corresponds to 4 to 6 weeks of maturation and at the beginning of puberty to 2 years. Girls are physiologically more mature in some other organ systems too and it seems very likely that this is the reason why more girls than boys survive at birth, whatever the general level of perinatal mortality.

The actual steps in the maturing process, the milestones passed, are the same until puberty for girls and boy.s They are also arranged in the same order of sequence along the road. But the intervals between some of them are different in the two sexes. The permanent teeth all erupt earlier in girls than in boys. However, some of them erupt much earlier, e.g. the canines with a difference of 11 months; others only slightly earlier, e.g. the first molars, with a difference of 2 months. Similarly, there are sex differences in the relative times of appearance of some centres of skeletal ossification, especially in the elbow region. Such differences must represent the small change of evolutionary advantage through some better cooperative arrangements for survival and for the rearing of young, but it is hard now to guess what they might have been. The earlier maturing of females is a characteristic shared by many mammals and nearly all primates so far examined. There are a few organs which constitute exceptions to the girls-earlier rule, most notably the milk teeth, which both sexes acquire at the same age.

Most of the sex differences in body shape and tissue composition are due to the events of puberty described in the next chapter. But the difference in overall body size is chiefly due to the boys' delay in growth. Though boys are slightly larger at birth than girls, the difference is small and remains so until the girls accelerate in their adolescent growth spurt. Girls then become temporarily taller. The boys, however, have meanwhile two years in which to continue their pre-adolescent growth before embarking on their own spurt. The actual amount of height gained in the spurt is, on average, greater in boys than in girls but only by some 3 to 5 cm. The height difference between men and women averages about 13 cm., of which 8 to 10 cm. is due to those extra pre-pubertal years of growth in boys.

The relatively longer legs of men compared with women (relative, that is, to total height) are due to the same mechanism. During the immediately pre-pubertal years the legs are growing relatively faster than the trunk. Thus during the pre-adolescent years boys change proportions to become longer-legged, and though during the spurt itself growth in trunk length is slightly greater in boys, it is not sufficient to cancel the effect of the pre-pubertal growth.

CHAPTER 5
Puberty

Puberty is the time of the greatest sex differentiation since the early intra-uterine months. There are changes in the reproductive organs and the secondary sex characters, in body size and shape, in the relative proportions of muscle, fat and bone, and in a variety of physiological functions. In the older literature the word puberty was used to denote the appearance of pubic hair, but nowadays it refers to the period at which the testes, prostate gland and seminal vesicles, or the uterus and vagina, suddenly enlarge. In this way it is possible to refer to puberty in mammals in general; in the few who reproduce these changes at each breeding season, 'puberty' refers to the first of such occasions. (The word adolescence is increasingly used to refer exclusively to the psychological and behavioural changes occurring around this time. However, such phrases as 'adolescent growth spurt' are very much part of the literature of human growth, and 'puberty' and 'adolescence' will be used interchangeably in this chapter.)

Reproductive Organs
   Males. Figure 21 shows the growth curve of the testes in healthy Swiss, Swedish and Dutch boys. The sizes of testes were determined by comparing them, by manual palpation, with standard models of increasing size. The models, for convenience assembled like a necklace, are known as the Prader orchidometer, after the originator, the Professor of Paediatrics in Zurich. Although the data are plotted against chronological age and thus subject to the difficulties described on page 11 above, they show the steep rise occurring at about 14 years and also the large differences between boys of the same age (as instanced by the gap between the 10th and 90th centile lines) that we shall discuss at length below.
   Beginning testicular enlargement is usually the first sign of puberty in boys, accompanied by changes in the texture and colour of the skin of the scrotum. A little later the penis starts to enlarge and the pubic hair to appear. The sequence of events is shown in diagram form in the lower part of Figure 22. The solid areas marked *testis* and *penis* represent the period of accelerated growth of these

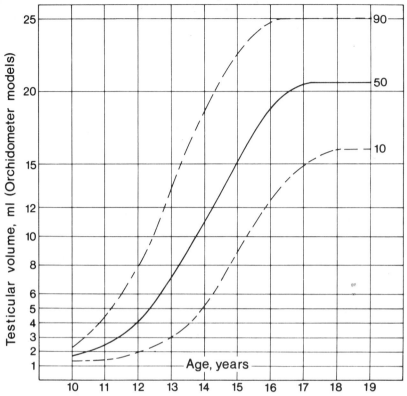

FIG. 21. Growth curve of testis size. Estimated 10th, 50th and 90th centiles, based on Dutch, Swiss and Swedish males
(Data from van Wiereingen *et al.*, 1971; Zachmann *et al.*, 1974; Taranger *et al.*, 1976)

organs, and the horizontal lines and rating numbers marked *pubic hair* stand for its advent and development. The sequence and timings represent average values for European and North American boys. To give an idea of the individual departures from average, the range of ages at which spurts for height, testis growth and penis growth begin and end are inserted underneath the first and last points of the curves or bars. The acceleration of penis growth, for example, begins on average at about age 12½, but sometimes as early as 10½ and sometimes as late as 14½. The completion of penis development usually occurs at about age 14½ but in some boys it is as early as 12½ and in others as late as 16½. There are a few boys therefore who do not begin this phase of pubertal development until the earliest maturers have completed it. Figure 23a shows three boys, all exactly the same age. At ages 13, 14 and 15 there is an enormous variability amongst any group of boys; this ineluctable fact of biology raises difficult social and educational problems, and contributes to the psychological maladjustments often seen in adolescents.

The *sequence* of events, though not exactly the same for each boy, is much less variable than the age at which the events occur. The spurt in height and other body dimensions begins on average about a year after the first testicular enlargement and reaches its maximum (peak height velocity) after about a further year, when the penis is also growing maximally and pubic hair is in stage 3 or 4.

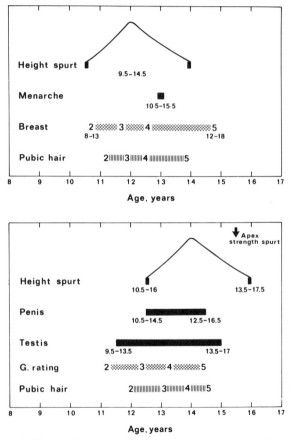

FIG. 22. Diagram of sequence of events at puberty in girls (*above*) and boys (*below*) (From Marshall and Tanner, 1970)

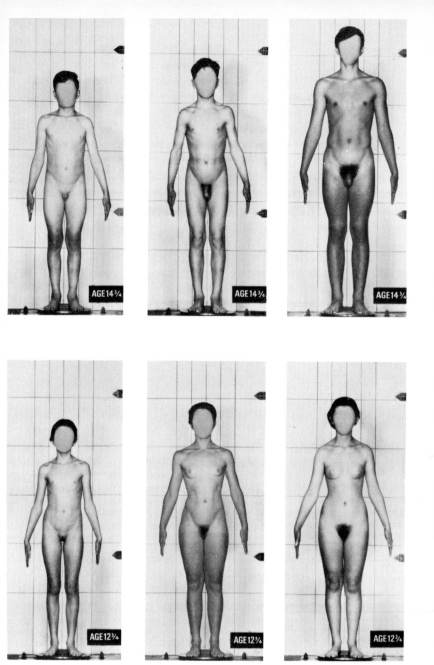

FIG. 23. Three boys, aged 14, at different stages of puberty; three girls, aged 12, at different stages of puberty
(From Tanner, 1975)

Axillary hair appears usually only when pubic hair has reached stage 4. Facial hair appears somewhat later, and usually in a definite order; first at the corners of the upper lip, then over the whole upper lip, then on the upper cheeks and the mid-line below the lower lip, and finally along the sides and lower border of the chin. The remainder of the body hair develops gradually from about the time of the first appearance of the moustache until a considerable time after puberty. The ultimate amount of general body hair that the individual develops seems largely to depend on heredity and is unrelated to the development of the specifically sexual hair on pubes and axilla.

Breaking of the voice happens relatively late in puberty and is due to the increased length of the vocal cords which follows the growth spurt of the larynx. However, since higher notes may still be made by correct stopping of the cords (as witness male countertenors), 'voice-breaking', which simply means the advent of lower tones due to enlargement of the larynx and lengthening of the vocal cords, is often a gradual process and is not reliable as a criterion of puberty (though it was often used as such at periods when examination of genitalia and pubic hair was regarded as improper).

During adolescence the breast in boys undergoes changes, some temporary and some permanent. The diameter of the areola, which is equal in both sexes before puberty, increases considerably, though not as much as in girls. The average diameter before puberty is about 12·5 mm. ($\frac{1}{2}$ in.) and this approximately doubles in boys and triples in girls. In some boys (between one-fifth and one-third of most groups studied) there is a distinct enlargement of the breast about midway through adolescence. When marked, this is known as gynaecomastia. The enlargement usually disappears spontaneously after about a year (if it does not, the tissue can be easily removed surgically).

Concurrently with growth of the penis, and under the same influence, the prostate gland and seminal vesicles enlarge. The secretion of the prostate makes up most of the seminal fluid. Thus no true ejaculation occurs before puberty, though the neural paths underlying orgasm are present. The first spontaneous ejaculation of seminal fluid usually occurs about a year after the beginning of penis growth, often during sleep and accompanied by appropriate dreaming. At first the seminal fluid seems to contain fewer and less viable sperm than in early adulthood; thus in males there may be an initial period of reduced fertility similar to that often occurring in girls (see below). In neither sex, however, can such relative infertility be relied on.

In clinical work some designation of how far a child has progressed through puberty is often required, and rating-scales have been developed for successive stages of growth of the penis and scrotum in the male, breasts in the female, and pubic hair in both sexes. The scales in each instance range from 1 (pre-pubescent) to 5 (adult). Details will be found in Tanner (1962), and illustrations in Chapter 11 (pp. 197 ff.). The size of the testes is assessed with the orchidometer. Genitalia and pubic hair do not necessarily develop in parallel; indeed some 15% of boys in the Harpenden Growth Study reached stage 4 of genitalia without pubic hair having yet appeared. The converse sequence, however, is exceedingly rare.

There are large individual differences in the rapidity with which a given sequence, once begun, progresses to maturity. For example, on average boys take about a year to go from the beginning of stage G2 to the beginning of stage G3 and about 3 years to go from G2 to G5. But some boys take only two years to go from G2 all the way to G5. Thus out of a collection of boys all starting equal at the beginning of G2 a few reach G5 before others even reach G3. This variation in speed of development is additional to the variation in age of puberty as a whole, and for the most part independent of it.

Females. The same great variability is seen in the development of girls at puberty. The upper portion of Figure 22 shows the average age at which each event takes place, and the range of variation in healthy girls. Usually the first event to be noticed is the advent of breast-stage 2, called the breast bud. This consists of an elevation of breast and papilla as a small mound, with slight enlargement of the areolar area. The average age at which this occurs in European and North American girls is about 11·0 years but the range extends from 9·0 to 13·0 (all such ranges given here refer to the mean plus and minus 2 standard deviations, which includes 95% of the normal population. Five per cent lie even beyond these limits). Usually pubic hair begins to appear a little later; however, in about one-third of all girls pubic hair appears before the breast bud. The two series of events show considerable independence; thus among girls in breast-stage 3 stages of pubic hair development are shown, 25% of girls having none at all, and 10% having reached full adult status.

Menarche, the first menstrual period (pronounced menárkee, from the Greek word archē meaning beginning) occurs relatively late in puberty. The average age of its occurrence varies from 12·8 to 13·2 years in various present-day populations of northern and central Europe and in girls of European descent in North America. There are extensive data on age at menarche from all over the world, shown in Table 3 (p. 143). The standard deviation of age at menarche is about 1 year in most populations so the 95% range is from about 11·0 years to about 15·0 years. Again there is considerable independence from other pubertal characteristics. Most girls at menarche are in breast-stage 4 (projection of the nipple and areola above the level of the rest of the breast, in contrast to stage 5 in which only the nipple projects, the areola receding to the general breast level). However, some 25% are in breast-stage 3, and a small percentage in the earlier, breast-bud stage. Similarly most girls at menarche are in pubic-hair stages 3 or 4, but some are in stage 5 and a very few still in stage 1. The relation of menarche to the height spurt, however, is close. All girls start to menstruate when the height velocity is falling, i.e. during the downward part of the height velocity curve. Indeed, on average menarche occurs at the time of maximum deceleration of height growth, the moment when the velocity is dropping fastest. The hormonal significance of this is unknown.

Though menarche marks a definitive and probably mature stage of uterine development, it does not signify the attainment of full reproductive function. The early menstrual cycles, which in some girls are more irregular than later ones, often take place without an egg being shed from the ovary. Thus there is frequently a period of adolescent sterility or partial sterility, lasting for 12 to 18 months after menarche. However, this does not occur in all girls.

There is an important difference between girls and boys in the relative positions of the height spurt in the whole sequence of development at puberty. It has only recently been realized, as a result of detailed longitudinal studies on relatively large numbers of children, that girls have their height spurt considerably earlier than boys. The difference between the ages of peak height velocity in girls and boys is just 2 years, but the difference in the first appearance of pubic hair is about 9 months. The first appearance of breasts precedes the first change in testis-size by even less. The adolescent growth spurt in girls is placed earlier in the sequence than in boys. Indeed in some girls peak height velocity is reached soon after breast-stage 2 and occasionally even before it. Thus for girls the first event of puberty is often an increase of growth velocity, seldom noticed. In boys, on the other hand, the peak height velocity very rarely occurs before stage 4 in genitalia growth is reached, and it is never the first sign of puberty. This has practical importance, for boys who are late maturers can be reassured that their height spurt is yet to come if genital development is not far advanced; and girls who worry about being too tall can be reassured their height spurt is nearly over if menarche has occurred. On average girls grow about 6 cm. (roughly 2 in.) after menarche, although gains of up to twice this amount may occur. The postmenarcheal gain is practically independent of whether menarche itself occurs early or late.

As in boys, some girls pass rapidly through all stages of breast or pubic hair development while others are slow. On average the time taken from first appearance of breast bud to reach breast-stage 3 is 1 year and to reach the adult stage is 4 years. Girls with rapid transit, however, take only $1\frac{1}{2}$ years to pass through all the stages, while slower developers take as much as 5 years or even more.

The Adolescent Growth Spurt: changes in body size and shape
At puberty there is a swift increase in body size, and a change in shape and in body composition. Though the changes are qualitatively similar for both boys and girls, they are quantitatively very different and their distinctive balance brings about much of the sexual dimorphism characteristic of adults. Thus before puberty males are only slightly taller than females whereas after it the difference averages 13 cm. (5 in.). At puberty the shoulders and muscles grow more in boys whereas the hips grow more in girls. Large differences in strength develop, together with other physiological distinctions. Some such sexual dimorphism occurs in all primates and indeed in most mammals, and relates to the behavioural as well as to the physical aspects of reproduction, and to the manner in which the species' social life is organized. The structural and physiological changes in boys make them more capable than girls of doing heavy physical work and of running faster and farther. Such sexual specializations are more notable in some species of primates than in others. Gibbons, for example, show little sex dimorphism; gorillas and rhesus monkeys much. Man lies about the middle of the primate range. Here again we have an ineluctable fact of biological history. The dimorphism may be no longer very relevant to many tasks in our push-button age but it is there nevertheless, with implications that cannot sensibly be denied or ignored.

The adolescent growth spurt in height is illustrated in the growth curves of the typical boy and girl given in Figure 5 (p. 14). The rate of growth falls continuously from birth onwards, and just before puberty it reaches its lowest point. In the year just preceding his spurt the average boy grows about 5·0 cm. and a gain of only 3·5 cm. is within the limits of normal. Thus boys with a delay of puberty who see others shooting up around them may feel they are not growing at all, so great is the contrast. Girls, with their earlier adolescent spurt, do not fall to quite so low a velocity before it. When the spurt starts, however, growth becomes very rapid. Typically a boy puts on about 7 cm. in the first year of the spurt, 9 cm. in the second and 7 cm. in the third. After this, growth rapidly diminishes, with 3 cm. in the next year and about 2 cm. thereafter. In girls the velocity is lower, at about 6, 8 and 6 cm. a year during the three years of the spurt. Naturally individuals differ much in the magnitude of their spurts; in Chapter 11 standard charts are given outlining the limits of normal. During the year which includes their peak velocities, however, most boys grow between 7 and 12 cm., and most girls between 6 and 11 cm. Children who have their peak

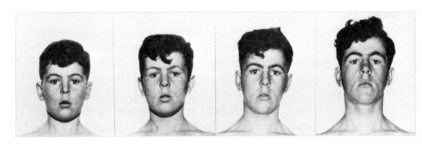

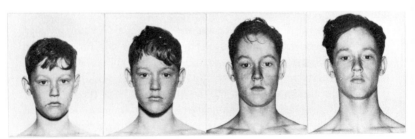

Fig. 24. Face changes in two boys at puberty
(rom Tanner, 1962)

early reach a higher peak than those who have it late; the early maturers are taking off from a higher level of velocity, though this is not the only factor involved.

The average age at which the peak is reached depends on the nature and circumstances of the population, but in European and North American children at present it is about 12·0 years for girls and 14·0 years for boys. The variation around this is about two years either way; thus very early maturing boys reach their peak at 12·0 years, having started their spurt at about 10·5, while very late maturers begin the spurt only at 14·5 years, and reach their peak about 16·0. Corresponding variation occurs in girls.

The adolescent spurt is under somewhat different hormonal control from growth in the preceding period (see Chapter 7) and probably as a consequence the amount of height added during the spurt is to a considerable degree independent of the amount attained before. Most children who have grown steadily up the average curve have an average adolescent spurt, but some end up considerably above average as adults, and others considerably below. In statistical terms, the correlation coefficient between adult height and height immediately before the spurt starts is 0·8. This means that about 30% of the variability among men in adult height is caused by differences in the magnitude of their adolescent growth spurts. The same is true of women.

More of the spurt in height comes from acceleration in trunk length than from acceleration in growth of the legs (in practice, trunk length is measured by sitting height, which thus includes head height and pelvis height; and leg length is stature less sitting height). There is a fairly regular order in which the dimensions accelerate; leg length as a rule reaches its peak first, some 6 to 9 months ahead of trunk length. Shoulder and chest breadths are the last to reach their peaks. Thus a boy stops growing out of his trousers (at least in length) a year before he stops growing out of his jackets. At adolescence some children, particularly girls, complain of having large hands and feet; at least they can be reassured that by the time the spurt has ended their hands and feet will be a little smaller in proportion to arms, legs and height.

Nearly all skeletal and muscular dimensions take part in the spurt, though not to an equal degree. Even the head diameters, practically dormant since a few years after birth, show a small acceleration in most persons. The skull bones increase some 15% in thickness and there is growth in the tissues of the scalp. It is doubtful whether the brain itself has an adolescent growth spurt; if it does, the extent of it is small. The face undergoes considerable changes, illustrated in Figure 24. The forehead becomes more prominent owing to growth of the brow ridges and the air sinuses behind them, and both jaws grow forward, the lower more than the upper. This change makes the profile straighter and the chin more pointed. At the same time the facial muscles develop. All these changes are considerably more marked in boys than in girls; indeed in some girls scarcely any detectable spurt in face dimensions occurs at all.

The greatest sexual dimorphism of the skeleton occurring at puberty, however, concerns the shoulders and hips. The curves are illustrated in Figure 25, aligned so that each individual's age is measured in years before and after his own peak height velocity. This minimizes the confusion caused by some individuals' having their spurt early and some late, and also places both sexes on the same time base. Girls have a particularly large adolescent spurt in hip width, a spurt which is quantitatively as great as that of the boys, despite the girls' spurt being a good deal less in nearly all other dimensions. The shoulder width spurt, on the other hand, is particularly marked in boys. These differences occur because cartilage cells in the hip joint are specialized to respond to female sex hormone (oestrogen) and cartilage cells in the shoulder region specialized to respond to male sex hormone (androgens, primarily testosterone). Such specializations can be lost as well as gained during the course of evolution. In most apes and monkeys the male has a large canine tooth, suitable for fighting. This normally develops at puberty, but it can be induced to grow before then by giving the animal testosterone. In man, the canine, though still slightly larger in the male, has apparently lost most of its ability to respond to androgens.

The shoulder–hip-width dimorphism has long been used as a measure of bodily androgyny, i.e. the degree to which the male resembles a female, or vice versa. In young adults an equation using just these two measurements (known technically as a discriminant function) classifies correctly 90% of persons into male and female. The score which does this is *3 Biacromal diameter–1 Bi-iliac diameter* and it may be used as a measure of androgyny which is less biased by sheer body size than is the simple ratio of shoulder hip widths.

Many sex dimorphisms antedate puberty, as pointed out in the

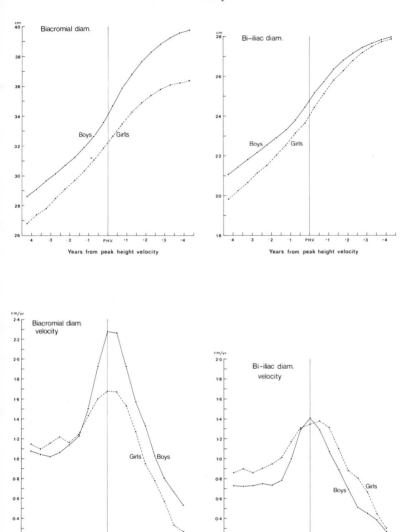

Fig. 25. Biacromial and bi-iliac diameters at puberty: distance (*above*) and velocity (*below*) curves. Longitudinal data of Harpenden Growth Study plotted in years before and after peak height velocity for each individual
(From Tanner, 1975)

preceding chapter. One of the most striking, the greater relative length of the male leg, comes about through males having a delayed adolescent spurt. Part of the sex difference in shape of the pelvis also antedates puberty. Girls at birth already have a wider pelvic outlet, i.e. a wider opening at the bottom of the bony pelvis, which constitutes the narrowest part of the passage through which the baby has to pass when it is born. Thus the adaptation for child-bearing is present from a very early age. The changes at puberty are concerned more with widening the pelvic inlet and broadening the hips. This is important in providing room for the uterus and baby to grow in.

Perhaps the hip widening may also have the subsidiary function of attracting the male's attention: for these sex-differentiated morphological characters arising at puberty – to which we may add the corresponding physiological and possibly psychological ones as well – are developed not only for immediate use in copulation, but for all the range of behaviour that accompanies sexual reproduction. The wide shoulders and muscular power of the male may have been developed partly for use in mating, but more probably, with the canine teeth and prominent brow ridges, for driving away other males as well as for hunting. The dimorphism is at first maintained by sexual selection, since fighting involves males of the same species. When this type of sexual selection becomes less important, as in man, genetical theory leads us to suppose that these traits are gradually eliminated.

But a number persist, through a mechanism known as ritualization. In the course of evolution a morphological character or a piece of behaviour may lose its original function and, becoming further elaborated, complicated or simplified, may serve as a sign-stimulus to other members of the same group, releasing behaviour that is in some way advantageous to the spread or survival of the species. It requires little insight into human erotics to suppose that the shoulders, the hips and buttocks, and the breasts (at least in a number of widespread cultures) serve as releasers of mating behaviour. The pubic hair (about whose function the medical text-books have always preserved a cautious silence) probably survives as a ritualized stimulus for sexual activity, developed by simplification from the hair remaining in the inguinal and axillary regions for the infant to cling to when still transported, as in many apes and monkeys, under the mother's body. Similar considerations may apply to the axillary hair, which is associated with special glands

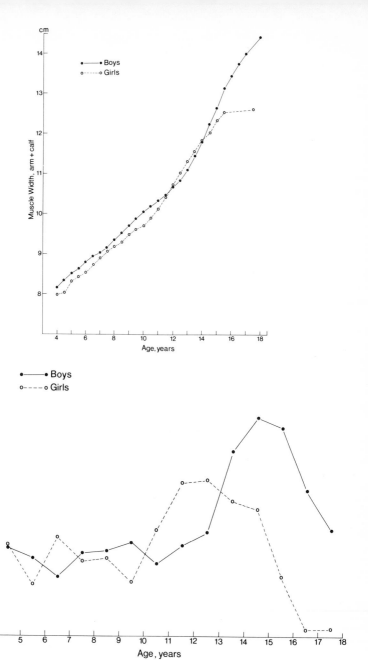

Fig. 26. Muscle growth at adolescence: sum of widths of upper arm and calf muscles as visualized in x-ray pictures: (*above*) muscle width attained; (*below*) velocity of muscle width (From Tanner, Hughes and Whitehouse, unpublished)

which themselves only develop at puberty and are related in structure to scent glands in other mammals. The beard, on the other
hand, may still be more frightening to other males than enticing to
females. But perhaps there are two sorts of beard.

*The Adolescent Growth Spurt: changes in proportions of muscle, fat
and bone*

There are marked changes in body composition accompanying the
adolescent spurt, especially in boys. Figure 26 shows the increase of
limb muscle, estimated from radiographs of arm and calf taken so

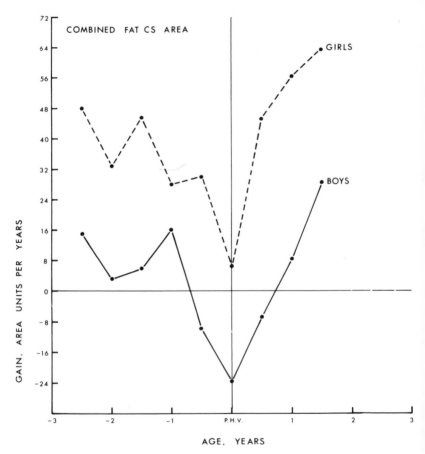

FIG. 27. Subcutaneous fat growth in upper arm and calf at adolescence, as visualized in x-ray
pictures
(From Tanner, 1975)

that bone, muscle and fat can be distinguished. The peak of the muscle growth velocity is somewhat later than the peak height velocity and thus occurs coincidentally with the peak of sitting height and shoulder width. It is much higher in boys than in girls. However, girls do have an adolescent spurt in muscle, and since it occurs earlier than in boys, girls on average actually have more muscle than boys for a short period, as Figure 26 shows. The increase in size of muscles reflects increased amounts of contractile protein and of nuclei, and underlies the great increase of strength occurring simultaneously (see below). Other muscles, including the heart, show similar growth curves.

Subcutaneous fat, or at least subcutaneous fat on the limbs, has quite the opposite curve. Figure 27 shows the velocity curve of limb fat in a group of children followed longitudinally, whose curves have been aligned as before, so that age is measured in years before and after their own peak height velocity. In boys the area of fat seen in the x-rays diminishes markedly, and the velocity actually becomes negative, indicating absolute loss. The lowest point is coincident with the peak of height velocity. In girls a similar thing occurs but is less marked, the decrease not being sufficient to carry the mean below zero. Thus no actual loss occurs in the average girl, who has to content herself with a temporary slowing down of fat accumulation. Fat on the body shows a much smaller decrease of velocity than fat on the limbs.

Strength, exercise tolerance and other physiological functions
The increase in muscle size is naturally accompanied by an increase in strength, illustrated in Figure 28. This is much greater in boys than in girls, and it seems probable that the strength per gramme of muscle becomes greater in boys at this time, owing to changes in the structural and biochemical nature of the muscle cells induced by male sex hormone. The curves in Figure 28 are from a longitudinal study. Each individual test represents the best of three trials, made with each child competing against his own previous reading, and also against a classmate of similar ability. Only with such precautions can maximal values be approached. Other data indicate that except in hands and forearms, girls and boys are similar in strength for given body size until puberty begins, when a wide gap opens even between persons of the same overall size.

Boys develop larger hearts as well as larger skeletal muscles, larger lungs, higher systolic blood pressure, lower resting heart-

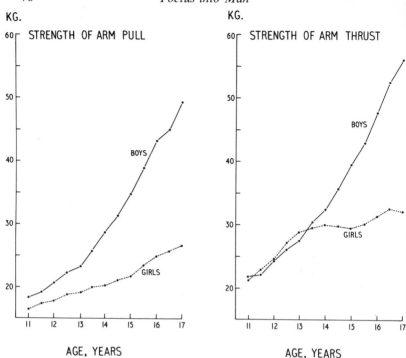

Fig. 28. Strength of arm pull and arm thrust from ages 11 to 17. Mixed longitudinal data, 65–93 boys and 66–93 girls in each group
(From Tanner, 1962, based on data of Jones)

rate, a greater capacity for carrying oxygen in the blood, and a greater power of neutralizing the chemical products of muscular exercise. The number of red blood cells and the amount of haemo-globin, the pigment which carries oxygen from lungs to muscles, both increase at adolescence in boys, but remain unchanged in girls. In short, the male becomes more adapted at puberty for the tasks of hunting, fighting and manipulating all sorts of heavy objects.

It is as a direct result of these changes that athletic ability increases so much in boys at adolescence. The popular notion of a boy 'outgrowing his strength' at this time has little scientific sup-port. On the contrary, power, athletic skill and physical endurance all increase progressively and rapidly throughout adolescence. If the adolescent becomes weak and easily exhausted, it is usually for psychological reasons, not physiological ones. However, there is a brief period during which trunk length has increased relatively to the legs, bringing new problems of balance, while the muscles have yet to reach their full size and strength. This period seldom lasts more than six months, but may bring temporary problems in such specialized groups as young male ballet dancers, weight-lifters and field-event athletes.

CHAPTER 6

Developmental Age, and the Problems of Early and Late Maturers

Puberty starts at very different ages in different individuals, even of the same sex. In the last chapter, in Figure 23, p. 63, we saw how perfectly normal 14-year-old boys or 12-year-old girls could be pre-pubertal, mid-pubertal or post-pubertal. Manifestly it is ridiculous to consider the three boys illustrated as equally grown up physically, psychologically or – since much behaviour at this age is conditioned by physical status – in their social relations. One simply should not talk of '14-year-olds': the statement that a boy is 14 is hopelessly vague, for so much depends on whether he is an early or a late maturer.

These differences in rate of maturation do not suddenly arise at adolescence, although they become more obvious then. They are present at all ages, even at birth. Just as the average girl is developmentally ahead of the average boy from the middle of the foetal period onwards, so the early-maturing boy, on average, is always ahead of the late-maturing. This tendency for development to be rapid or slow was aptly named *tempo of growth* by Franz Boas, one of the great pioneers of human auxology. Some children play out their growth andante, others allegro, a few lentissimo. It seems that heredity plays a large part in setting the metronome, but we do not know the physiological mechanism.

Evidently we need a measure of developmental age, or physiological maturity, which represents more truthfully than chronological age how far a given individual has progressed along his or her road to full maturity. Height is not such a measure, because persons differ in mature height. Thus tallness in a child may signify either a rapid tempo of growth in a child going to be of average height when adult, or an average tempo of growth in a child going to be tall when adult. The percentage of the individual's own mature height reached at any age is a better measure, but available only retrospectively. Age at entry to the various stages of puberty is a valid measure, since all normal children pass through the same stages; but this measure also is available only late in growth. However, there are many possible measures of developmental age, ranging from the number of erupted teeth to the percentage of water in muscle cells. The various 'age' scales do not necessarily coincide and each has its particular use. The measure most generally used derives from the successive stages of development of the skeleton as seen in radiographs. This measure is applicable throughout the whole period of growth and is called *skeletal maturity* or *bone age*.

Skeletal maturity is a measure of how far the bones of an area have progressed towards maturity, not in size, but in shape and in their relative positions one to another, as visualized in a radiograph. Each bone begins as a primary centre of ossification, passes through various stages of enlargement and shaping of the ossified area, acquires, in many cases, one or more epiphyses, i.e. other centres whose ossification begins independently of the main centre, and finally reaches adult form when these epiphyses fuse with the main body of the bone (see pp. 32–3). All these changes can be easily seen in radiographs, which distinguish the ossified area, whose calcium content renders it opaque to the x-rays. The sequence of changes of shape through which each bone passes is the same in all individuals, and all individuals reach the same final state.

In principle, any or all parts of the skeleton could be used, but in practice the hand and wrist forms the most convenient area. It is easily x-rayed without fear of any radiation being delivered to the reproductive organs, and it requires only a minute dose of x-rays, of the order of 4 millirads. There is no evidence that such a dose does any harm whatever. Indeed it would be surprising if it did, since it represents the dose obtained unavoidably by every child who spends a week on holiday in the mountains (background radiation is 100 millirads per year at sea level, rising to 300 millirads at 2,000 metres altitude). The public is justifiably worried about the medical and genetic effects of atomic fallout and about the doses of radiation given in some forms of diagnostic radiology, but it is necessary to think quantitatively to keep matters in perspective.

Figure 29 shows two radiographs, both of boys aged 14·0 years. The one on the left has many epiphyses still unfused, and the wrist bones are not yet fully shaped. The hand on the right is approaching maturity. Skeletal maturity is assessed either by the 'atlas' bone age method of Greulich and Pyle or by the more detailed and perhaps more accurate bone-scoring method of Tanner and Whitehouse. In the former each bone is matched with a similar-appearing bone in a series of standard radiographs of increasing

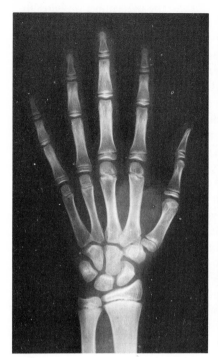
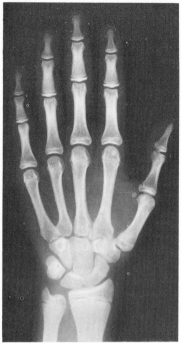

FIG. 29. Radiographs of two boys both aged 14·0 years: (*left*) bone age 12·0 'years'; (*right*) bone age 16·0 'years'

bone age. Thus each bone has a bone age assigned to it, and the modal (or most frequent) of these bone ages is taken as the bone age of the hand and wrist. Frequently the step of assigning bone ages to each separate bone is omitted; instead, the overall appearance of the given radiograph is simply compared with the appearance of the standard radiograph, and the nearest overall match is taken. But such short-circuiting all too easily leads to inaccurate work.

In the Tanner–Whitehouse method each bone in the given radiograph is matched with a set of 8 standard stages and its stage is rated in just the same way as the puberty stages. Short-circuiting is impossible. Each stage of each bone has a score attached to it, derived mathematically so that the sum of the scores for all the bones represents the best overall estimate of skeletal maturity. In this way, a score for skeletal maturity is obtained which is comparable with a measure of height except that it runs from 1 to 100. The maturity score may be turned into a bone age if desired, the bone age being simply the chronological age for which the given score is at the 50th centile. The Tanner–Whitehouse standards are based on a large-scale random sample of Scottish urban and rural children taken in the 1950s, and the statement that a certain boy has a bone age of 8·0 'years' means that he has the same degree of skeletal maturity as the average 8·0-year-old Scottish boy of the area, social class and time. The Greulich–Pyle standards are based on a smaller number of very well-off North American children in the 1930s, who were some 6 to 9 months ahead of the Scottish children in maturity. Thus the two methods do not lead to the same result unless a correction is made. Workers in different countries or geographical areas have to establish local norms for skeletal maturity just as they do for height (see Chapter 9).

Boys and girls pass through identical stages in the Tanner–Whitehouse system and hence are rated in the same way. However, the scores attached to each of the stages are somewhat different, reflecting a sex difference in manner of maturation. The total maturity-score, obtained by summing the separate stage-scores, is, of course, much higher in girls than in boys of the same chronological age. Girls are advanced compared with boys from before birth till maturity, the bone age of boys averaging about 80% that of girls.

The hand and wrist provides useful information only from about 18 months onwards, so other parts of the skeleton, chiefly the legs and feet, have to be used from birth to about 18 months. For some purposes not all the bones of the hand and wrist need to be scored; the forearm bones (radius and ulna) and the finger bones (metacarpals and phalanges) suffice. This is the case when bone age is used to predict the adult height that a child will reach. At a given chronological age an advanced child will be nearer adult height than a delayed child, and allowing for this makes prediction more accurate. Tables for prediction, with and without the use of bone age, are given in *Assessment of Skeletal Maturity and Prediction of Adult Height* (see Tanner *et al.*, 1975). Dr Alex Roche, at the Fels Research Institute, together with statistical collaborators, has provided a method for assessing the knee for skeletal maturity, and shown that hand and knee may not always give a closely similar result, even in healthy children.

Dental Maturity

The age of eruption of the teeth may also be taken as a measure of maturity. The deciduous dentition erupts from 6 months to 2 years of age and can be used during that period. The permanent dentition provides a measure from about 6 to 13 years. From ages 2 to 6 and from 13 onwards little information is obtainable from simple counting of the number of teeth erupted, but stages of calcification of teeth as seen in jaw radiographs can be used for dental maturity in exactly the same way as stages of the bones of the hand and wrist (see Demerjian *et al.*, 1973).

Relations between different measures of maturity

Skeletal maturity and dental maturity are not closely related. It is true that from 6 to 13 years children who are advanced skeletally have on average more erupted teeth than those who are skeletally delayed. Likewise children who enter puberty early erupt their teeth early on average, and girls have on average more erupted teeth than boys. Such average relationships, however, have little application to the individual. The correlation between bone age and dental age at a given chronological age is of the order of 0·4. This means that the two systems, though reflecting faintly the underlying presence of an overall general factor of maturity, measure to a great extent separate sorts of maturation. This is scarcely surprising, considering how very different are the timings of the growth curves of the head and its associated structures and the growth of the rest of the skeleton.

Skeletal maturity is more closely related to the age at which puberty occurs, though even this relationship is less close than was formerly thought, at least within the range of normal children. The milestones that are most closely linked are age at menarche and bone age. Figure 30 shows the frequency distribution of age of menarche in relation to chronological age and skeletal age in the same, normal, girls. Evidently the variation in relation to bone age is much less; indeed the 95% ranges are 11·0 to 15·0 in terms of chronological age but only 12·0 to 14·0 in terms of bone age.

The link between skeletal maturity and the beginning of puberty (i.e. the appearance of breasts or the enlargement of testes) is looser. However, at the extremes of advancement or delay in normal children, and even more in the abnormally advanced or delayed, skeletal maturity is a good guide to what is happening and what is likely to happen. If a boy is very much delayed in puberty then estimation of his bone age provides important information. If

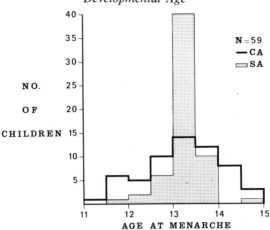

Fig. 30. Distribution of age of menarche in terms of chronological age and skeletal age. Girls of Harpenden Growth Study
(From Marshall, 1974)

bone age is delayed then development is all of a piece and we may expect puberty to occur in the course of time; he simply has delay in growth. If the bone age is at an immediately pre-pubertal level of about 14·5 'years' and remains so over a period, then something is actually stopping puberty from occurring. The way in which the initiation of puberty is controlled is discussed in the next chapter.

Early and late maturing, mental ability and emotional development
Clearly the occurrence of differences of tempo in growth may have profound implications for educational theory and practice, especially if advancement in physical growth is linked to any significant degree with advancement in intellectual ability and emotional maturity.

There is good evidence that, in European and North American school systems, children who are physically advanced towards maturity score on average slightly higher in most tests of mental ability than children of the same age who are physically less mature. The difference is not great, but it is consistent, and it occurs at all ages that have been studied, going back as far as 6·5 years. Similarly, the intelligence test score of postmenarcheal girls is higher than the score of premenarcheal girls of the same age. Thus in age-linked examinations physically fast-maturing children have a significantly better chance than slow-maturing children.

It is also true that physically large children score higher than small ones, at all ages from 6 years on. In a random sample of all Scottish 11-year-old children, comprising 6,940 pupils, the correlation between height and score in the Moray House group test of ability was $0·25 \pm 0·01$, allowing for the effect of age differences within the range $11·0$ to $11·9$ years. An approximate conversion of these scores to Terman–Merrill I.Q. leads to an average increase of $0·67$ points for each centimetre of stature, or roughly $1\frac{1}{2}$ points per inch. A similar correlation has been found in London children. The effects can be very significant for individual children. In 10-year-old girls there was a 9-point difference in I.Q. between those whose height was above the 75th centile and those whose height was below the 15th. This is two-thirds of the standard deviation of the test score.

Children with many siblings (brothers and sisters) are shorter in stature and score less in intelligence tests than children with few sibs. About half of the correlation in the paragraph above is associated with differences in numbers of sibs; but about half remains when the number of sibs is allowed for.

It used to be thought that both the relationship between test score and height and between test score and early maturing would disappear in adulthood. If the correlations represented only the effects of co-advancement both of mental ability and physical growth, this would be expected to happen. There is indeed no difference in height between early- and late-maturing boys when both have finished growing. But it is now clear that, curiously, at least part of the height–I.Q. correlation persists in adults. It is not clear in what proportion genetical and environmental factors are responsible; differential social mobility (i.e. children's rising or falling in the social hierarchy relative to their parents) is probably the main factor involved (see also Chapter 9, pp. 146–9).

There is little doubt that being an early or a late maturer may have repercussions on behaviour and that in some children these repercussions may be considerable. There is little enough solid information on the relation between emotional and physiological development, but what there is supports the commonsense notion that emotional attitudes are clearly related to physiological events.

The world of the small boy is one where physical prowess brings prestige as well as success, where the body is very much an instrument of the person. Boys who are advanced in development, not only at puberty but before as well, are more likely than others to be leaders. Indeed this is reinforced by the fact that muscular,

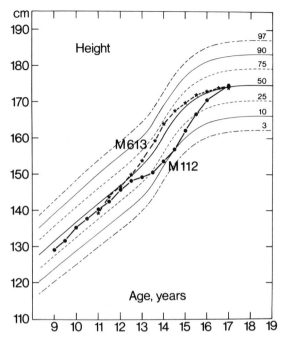

Fig. 31. Height attained from ages 11 to 17 of two boys of the Harpenden Growth Study. One (M613) has an early, and the other (M112) a later, adolescent spurt. The plots are made against the Tanner–Whitehouse 1975 British Standards.
(Redrawn from Tanner, 1977)

powerful boys on average mature earlier than others and have an early adolescent growth spurt. The athletically built boy not only tends to dominate his fellows before puberty, but by getting an early start is in a good position to continue that domination. The unathletic, lanky boy, unable, perhaps, to hold his own in the pre-adolescent rough and tumble, gets still further pushed to the wall at adolescence as he sees others shoot up while he remains nearly stationary in growth. Even boys several years younger now suddenly surpass him in size and athletic skill and perhaps too in social grace. Figure 31 shows the height curves of two boys, the first an early-maturing, muscular boy, the other a late-maturing, lanky boy. Although both boys are of average height at age 11, and then are together again at average height at age 17, the early maturer is 12 cm. taller during the peak of adolescence.

The late developing girl, though not perhaps so plagued by athletic problems, may have particular difficulty in her social relations because her lack of breasts is plain for all to see. Indeed, late developers at adolescence may sometimes have doubts about whether they will ever develop their bodies properly and whether they will be as well endowed sexually as others seen developing around them. At a deeper level the lack of development may act as a trigger to reverberate fears accumulated deeper in the mind during the early years of life.

It may seem as though early maturers have things all their own way. But early maturers have their difficulties also, particularly the girls in some societies. Though some glory in their new possessions, others are embarrassed by them. The early maturer has a longer period of frustration of sex drive, and of drive towards independence and the establishment of vocational orientation.

Studies by Dr Paul Mussen at the University of California have given evidence of the psychological consequences of early and late maturing. Late-maturing boys showed more attention-getting behaviour and were rated by their age peers and trained observers as more restless, talkative and bossy. They were less popular and had lower social status. In contrast, the outstanding leaders came from the early-maturing group. Projective techniques revealed in the late maturers deeper feelings of inadequacy, greater anxiety and greater anticipation of rejection by the group. Follow-up studies in young adulthood showed that, at least in the society of the United States, the early-maturing boys became more sociable and less neurotic.

Little can be done to diminish the individual differences in children's tempo of growth, for they are biologically rooted and not significantly reducible by any social action. It therefore behoves all teachers, psychologists and doctors to be both fully aware of the facts and alert to the individual problems they raise.

CHAPTER 7
The Endocrinology of Growth

The endocrine glands are collections of cells which manufacture chemical agents and discharge them into the blood-stream. The agents are known as hormones. Strictly speaking, a hormone is any substance secreted by one cell which passes to and has an action on another. Some hormones act on cells very close to their origin and do not enter the blood; some enter the blood but only in a special portion of the circulation designed to carry them a short distance to their target; and some enter the general blood-stream and are carried in it all over the body. Thus hormones constitute one of the two great systems of communication in the body (the nervous system being the other). This is as true for the carrying over of messages from one time to another as it is for the carrying of messages from place to place. The endocrine system is one of the chief agents for translating the instructions of the genes into the reality of the adult form, at the pace, and with the result, permitted by the available environment.

The hormones that are released into the general circulation pass by all manner of cells, but their targets are highly specific. This specificity is brought about by the existence of particular receptors in the target cells. The receptors are protein molecules; each abstracts its particular hormone from the blood, and links to it. Some receptor molecules are situated on the outer membrane of the cell, and others within the cell, in the cytoplasm. Those on the membranes react with hormones which are too large to be able to diffuse freely in and out of cells, as small molecules can. These hormones have to be grabbed from the blood (or more accurately from the tissue fluid, a filtrate of blood) as it flows by the cells. They are mostly peptides (strings of amino-acids linked together, in the same way as in proteins, which are essentially just very large peptides). The hormones from the pituitary gland (see pp. 89–91) are peptides of this sort. When the hormone and receptor link up on the membrane an intracellular messenger is released which speeds up a particular piece of cell machinery, in many cases by entering the cell nucleus and acting on the DNA of the chromosomes themselves. For many of these different hormone–receptor complexes the intracellular messenger is the same substance, called cyclic adenosinemonophosphate (cyclic A M P).

Smaller hormones pass back and forth through the cell membranes as a matter of course and are to be found in the cytoplasm of virtually all cells in the body, in concentrations reflecting their current levels in the blood. Their presence has no effect except on the particular cells which contain receptors. Receptors for these hormones are placed in the cytoplasm, and the link-up occurs there. The linked complex then itself travels into the nucleus and turns on a particular part of the D N A mechanism for manufacturing protein. Recently even the fine details of this intricate process have become clear (see O'Mally and Schrader, 1976). Steroid hormones, which include testosterone and oestradiol as well as cortisol (see pp. 95 ff. below) act in this way. Thus in the muscle cells of the uterus there are many receptors for oestradiol, the female sex hormone. When blood oestradiol rises, the receptor picks up more molecules and the complex causes new muscle protein to be made: the uterus grows.

Thus receptors are as important in endocrinology as the hormones themselves. Indeed it seems that the evolution of this system of communication is more a story of receptor development than of changes in the forms of the hormones too, though these have occurred as species evolved (see, for example, growth hormone, pp. 91–3). A particular receptor may of course be present in all sorts of different tissues scattered all over the body; hence the ability of testosterone to act on cells in the penis, muscles, shoulder cartilage and brain. Different tissues contain different amounts. It has been calculated that 80% of cells in the uterus (at least of rats) contain receptors for oestradiol, at about 12,000 molecules per cell, while in the hypothalamus only half the cells have oestrogen receptors, and at the lower level of 4,000 molecules per cell. The hormone–receptor mechanism is extremely sensitive: O'Mally and Schrader use a striking simile: 'If the palate were as sensitive to flavours as the target cells are to hormone', they write, 'we would be able to detect a pinch of sugar dissolved in a swimming pool.' Receptors allow for a certain complication of action also. Hormone A may act by altering the sensitivity of receptor B^1 to hormone B. If it does so it will be hard to distinguish this event from the straightforward occurrence of increase in hormone B. It is even believed that in some instances a hormone sensitizes its own receptor to its own action, a potentially explosive situation.

There are some dozen hormones which are of particular importance in the control of growth; the actions of each will be described below. They are thyroxine from the thyroid gland; cortisol and adrenal androgens from the cortex of the adrenal gland; testosterone from the Leydig cells of the testis; oestrogens from the ovary; insulin from the islets of Langerhans of the pancreas; and a series of hormones from the pituitary gland. These are growth hormone (GH); thyroid-stimulating hormone (TSH), also called thyrotrophin; adrenocorticotrophin (ACTH); and the gonadotrophic hormones, which are follicle-stimulating hormone (FSH) which stimulates growth of the young ovarian eggs and follicles of the ovary and the sperm-producing cells of the testis, and luteinizing hormone (LH) or Leydig-cell-stimulating hormone, which stimulates the growth and secretion of the corpus luteum in the ovary and the Leydig cells in the testis. The pituitary also produces prolactin, necessary for lactation, and vasopressin, which is necessary for control of body water. Another gonadotrophic hormone with actions similar but not identical to those of LH is produced by the cells of the chorion, which is the cell layer of the placenta nearest the maternal circulation. Chorionic gonadotrophin (CGT) enters primarily the maternal circulation, though it crosses from here into the foetus. The placenta also produces a hormone somewhat resembling growth hormone, called placental lactogen.

Control of Endocrine Secretion

Most of the endocrine glands secrete their hormones in response to messages, themselves hormonal, arriving from the pituitary gland. Thus the Leydig cells of the testis are stimulated to action by pituitary LH, the thyroid cells by pituitary TSH and so on. The pituitary, however (once known to generation upon generation of medical students as the 'conductor of the endocrine orchestra'), is only the manufacturing site and storage depot of these stimulus hormones. Their control resides in the hypothalamus, a portion of the brain geographically close to the pituitary and connected to it by a special system of blood vessels (see Figure 32). These are designed specifically to receive chemical messengers emanating from the tips of nerves in the hypothalamus and to pass them on to the cell receptors in the near-by pituitary. In the midst of this highly organized complexity it comes as a welcome surprise to learn that the hypothalamic nerve-secreted hormones are extraordinarily simple substances; so much so, in fact, that they can be synthesized on a commercial scale. The simplest of all is thyrotrophin-releasing

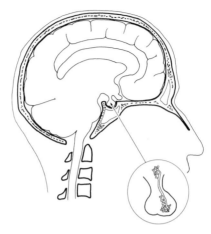

Fig. 32. Diagram showing position of pituitary gland immediately below, and joined to, the brain. The hypothalamus is the area of brain just above the pituitary. The circled inset shows the system of blood vessels going from the hypothalamus to the anterior part of the pituitary.

hormone (TRH). This is a tri-peptide, i.e. a molecule consisting of only 3 amino-acids. In contrast, human growth hormone (HGH), a fairly typical pituitary hormone from this point of view, consists of 191 amino-acids (though not, of course, all different ones). It seems a general rule that nerve cells can only synthesize and secrete rather simple substances, which serve as triggers to set off the larger, more complicated, charges manufactured in the endocrine glands.

Each pituitary hormone awaits its own specific hypothalamic message; TSH is released by TRH, ACTH by corticotrophin releaser. It seems that LHRH, luteinizing-hormone–releasing-hormone, a 10-amino-acid peptide, may be an exception, in that it releases FSH also, though to a smaller extent. No FSH releaser has yet been found. In addition two hormones, prolactin and growth hormone (which are of very similar structure and clearly derived from a common ancestral molecule in evolution), are controlled by being stopped instead of started. There is no known releasing hormone for either, but each has an inhibiting hormone which in the case of growth hormone is called somatostatin, consists of 14 amino-acids, and is available from commercial synthetic sources.

The hypothalamic releasing (or inhibiting) substances are themselves switched on or switched off in response to certain defined stimuli, of which the chief is the level of the relevant hormone circulating in the general blood-stream. This level is sensed by specialized cells in the hypothalamus near to the cells which synthesize and secrete the releasers. Thus the feedback circuit illustrated in Figure 32 is established: sensor ⟶ releaser ⟶ pituitary hormone ⟶ peripheral hormone ⟶ sensor. Outside influences can play on the circuit, particularly by altering the sensitivity of the sensor to circulating hormone. The way puberty is initiated provides an example. Before puberty the sensor for oestrogen is set so that it responds to a rise in blood oestrogen, by turning off the FSH-releasing hormone, when blood oestrogen is still at a very low level. Then impulses from other parts of the brain arrive which diminish the sensitivity of the sensor so that now all it responds to is a much higher level of oestrogen. Consequently, FSH and oestrogen rise until the new, pubertal, level of oestrogen is established.

There may be a number of sensors serving a single releaser. Furthermore, these sensors may change their relative importance at different times of life or in different circumstances. It seems, for instance, that in primates, just before menstruation occurs, the sensor serving FSH- and LH-releaser stops being turned off by oestrogen and is suddenly switched to being turned on by it (see further pp. 98–9 below). Hypothalamic sensors and releasers are still a rapidly developing area of knowledge, and there are certainly many sensors and probably some releasers of which we are still ignorant.

In certain circumstances, some hormones are released without the full circuit being involved. There is evidence that a rise in the level of blood thyroxine sometimes is sensed directly at pituitary level, thus switching off TSH immediately, without the intervention of TRH.

Growth Hormone (GH)

Growth hormone is necessary for normal growth from birth to adulthood. Children who lack it end up as adults about 130 cm. tall, but with normal proportions; they have been called miniatures. Growth hormone can now be extracted from human pituitary glands and given to such children by injection; they then grow at a normal rate, or a faster rate than normal if they are already small when the injections are started. (This is the phenomenon called catch-up growth, discussed further in Chapter 10.) Figure 33 shows

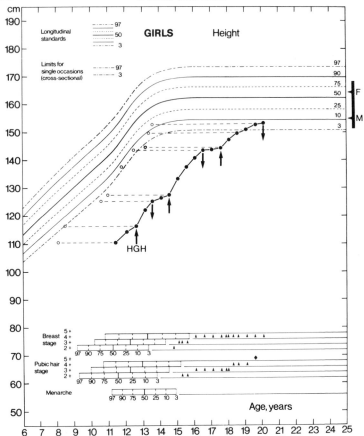

FIG. 33. Effects of growth-hormone deficiency on growth, with catch-up following treatment with human growth hormone (HGH) given over three periods indicated by arrows. Girl with isolated growth-hormone deficiency. Open circles represent bone age. F and M are father's and mother's height centiles, and bar represents predicted height centile range for their children. Puberty ratings at each age shown by arrowheads, lower section

the growth curve of a girl whose treatment started at 12·6 years. A further example is shown in Figure 78, p. 208.

Growth hormone has the curious property that it is species-specific. Whereas insulin and ACTH from animals works perfectly well in humans, GH from animals does not. Thus human growth hormone extracted at autopsy must be used, since with 191 amino-acids GH is too large a molecule to be synthesized commercially. Fortunately the pituitary glands of adults contain just as much GH as those of children.

See 100

GH is not necessary during foetal life, and it is not at all clear what it does in adults. Adults with untreated GH deficiency are perfectly healthy, and have no complaints apart from their small size. It is possible that the secretion rate diminishes after age 30 or so, despite the large quantities remaining in the pituitary, where the concentration is 1,000 times that of any of the other pituitary hormones.

Most pituitary hormones are secreted not continuously, but in a pulsatile fashion, in bursts. GH is a prime example of this. Most of the time the amounts of GH in the blood of children as well as adults are so low as to be scarcely detectable, but a few times in each 24 hours the levels rise for periods of the order of 30 or 60 minutes. GH is regularly secreted some 60 to 90 minutes after sleep commences, and usually in response to exercise also, and to anxiety. The mode of action of GH explains why this merely episodic release is sufficient. GH does not act directly on the bones to make them grow (as was thought for a long time), but on the liver to stimulate production of another hormone, called somatomedin. It is somatomedin, a smaller molecule, that acts on the growing cartilage cells at the ends of bones, and probably on muscle cells, whose growth is also stimulated by GH. It seems that a single pulse of GH suffices to raise blood somatomedin levels for at least 24 hours, thus ensuring continuous stimulation of growth. This is a rapidly-developing area of knowledge at the time of writing, but it seems unlikely that the outline above will substantially change as we discover more. One thing is tolerably clear: differences between normally large and normally small children and adults are not caused by differences in GH secretion nor, probably, by differences in somatomedin levels. Perhaps it is the receptors in the cartilage cells which control size. At any rate normal small children have plenty of GH and are not turned into normal big ones by being given GH in excess. Probably giving excess amounts of somatomedin, when it becomes available, will also fail to increase a growth rate that is within normal limits. In healthy children blood levels of somatomedin do not diminish with age to parallel the velocity curve, but, surprisingly, increase from infancy to puberty. Thus the normal velocity of growth depends on a more elaborate level of control.

Thyroid Hormone
Thyroxine, a small molecule available synthetically, is necessary for

normal growth from early foetal life onwards, and for normal physiological function in children and adults. It begins to be secreted at about the 15th to 20th postmenstrual week and is essential in the foetus and very young child for protein synthesis in the brain and for the proper development of nerve cells. As the brain matures this action becomes less prominent. Thus children born with thyroid deficiency become mentally handicapped unless treated at once. Diagnosis from clinical appearance is difficult in the first crucial days after birth, but routine biochemical screening of all newborns is technically possible and has been initiated in a few areas. The incidence of the disorder is probably about 1 in 5,000 births, which

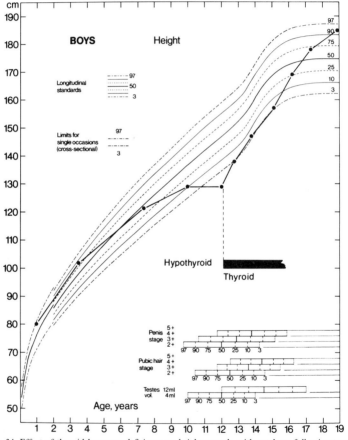

FIG. 34. Effect of thyroid hormone deficiency on height growth, with catch-up following treatment with thyroxine. Puberty ratings not available
(Redrawn from Prader, Tanner and von Harnack, 1963)

is similar to that of phenylketonuria (PKU), long screened with success. Treatment is very effective, so such screening seems well worthwhile.

Children who develop a deficiency of thyroid hormone at a later age have increasingly slow bodily growth and delayed skeletal and dental maturity, but do not suffer any particular brain damage. On treatment with thyroxine, such children show a great catch-up in growth and eventually reach perfectly normal adult height provided they are treated with reasonable promptness (Figure 34). It does not seem that the intensity of normal growth is regulated by the level of thyroid hormone any more than by the level of growth hormone. The thyroid's action is termed 'permissive'; that is to say, a normal thyroxine level permits the cells of the body to function properly; when there is too little, most cells, including those of the pituitary, function too slowly to be effective.

The rate of secretion of thyroxine is regulated by pituitary TSH (thyroid-stimulating hormone) itself controlled by hypothalamic TRH. Thyroxine level itself regulates TRH secretion, in a classical feedback loop.

Adrenal Hormones

The adrenal gland consists of an inner core, the medulla, and an outer shell, the cortex. The two are really quite separate glands from a functional point of view, and the medulla, which secretes mostly adrenalin, has little to do with growth.

The adrenal cortex secretes three groups of hormones:

1. **Mineralocorticoids.** These are aldosterone and to a smaller extent 11-deoxycorticosterone, and they maintain within acceptable limits the levels of sodium and potassium in the tissue fluids. Aldosterone secretion is not controlled by the pituitary, but by a local feedback, reacting to changes in blood composition and volume. Aldosterone is essential to life, but otherwise has no direct bearing on growth.

2. **Glucocorticoids** (or corticoids). Cortisol (also called hydrocortisone) and, to a lesser extent, corticosterone, are hormones secreted in a pulsatile manner throughout life, with a diurnal rhythm giving, under normal conditions, a maximum around noon and a minimum at about 0400 hours. Cortisol increases the formation of glucose from protein and has an anti-inflammatory and anti-stress action. The rate of secretion is raised to 2 or 3 times the ordinary level in response to the stresses of infection, extreme exer-

cise or severe emotion. Secretion is controlled by ACTH from the pituitary, and ACTH is itself released by the action of corticotrophin-releasing factor from the hypothalamus. The administration of (synthetic) cortisol, or its near relative cortisone, to persons whose pituitaries have been destroyed is essential if they are to lead anything approaching normal lives.

Cortisol is secreted during foetal life and throughout childhood in much the same amounts as in adults, relative to body size. No particular increase occurs at puberty, apart from that expected from the child's increased size. However, cortisol does have an anti-growth action. Though in normal amounts it probably plays no part in controlling rate of growth, if given in excess of normal it slows up growth in height and retards skeletal maturity, while at the same time causing increase in fat. This may easily occur in children treated with corticoids for severe asthma, rheumatoid arthritis (Still's disease), kidney disease, or severe eczema. In all these diseases corticoids should be avoided if possible; but sometimes they are absolutely necessary. There is evidence that giving them every other day instead of every day has considerably less effect on growth and often suffices therapeutically. Cortisol appears to block the ability of growth hormone to produce somatomedin; and this cannot be overcome by giving excess growth hormone. Cortisol may also directly inhibit the action of somatomedin on cartilage cells and perhaps the secretion of growth hormone itself.

See 100 3. **Androgens.** These are three or four closely-related substances secreted by the adrenal whose action is in most respects similar to that of testosterone, the male sex hormone. They are therefore called androgens ('andros' being Greek for man). Their function appears to be to cause some of the changes of puberty and perhaps to maintain the presence and size of some secondary sex characters, including muscle bulk. During childhood their rate of secretion is very low, but a marked increase takes place at puberty, to levels which are somewhat higher in men than in women. The highest levels in the blood are reached in early adulthood; thereafter the amount declines and by ages 60 to 70 returns to pre-pubertal values. The main hormone of this group is dehydroepiandrosterone.

Presumably these hormones are largely responsible in girls for the adolescent growth spurt and for the growth of pubic and axillary hair. They may also play some part in controlling maturation of the skeleton, i.e. the advance of bone age. In boys all these things are done, and more effectively, by the subjects' testosterone but adrenal androgens may supplement the action. The way in which adrenal androgen secretion is regulated is still something of a mystery. It is stimulated a little by ACTH but to a smaller extent than is cortisol secretion. The pubertal rise cannot simply be attributed to increase in ACTH, since cortisol does not rise concurrently. A special adrenal androgen-stimulating hormone has been postulated but never found.

Foetal Adrenal Cortex

During the foetal period there is a special zone of cells in the adrenal gland, lying between the interior medulla and the usual, exterior cortex. The zone appears at about 12 postmenstrual weeks, enlarges till birth or a little before, then swiftly atrophies, disappearing completely. At its maximum it occupies 80% of the whole adrenal cortex. It is dependent on and stimulated by a special pituitary hormone, only recently discovered and only present in the foetus, called CLIP, which stands for corticotrophin-like intermediate-lobe peptide. It secretes exclusively dehydroepiandrosterone and its sulphate (i.e. the androgens of the adult adrenal). These substances go to the placenta, where they are converted to oestrogens, which seem to be necessary in considerable quantity for the normal development of pregnancy.

In this, as in some other interactions, foetus and placenta function as a co-operative unit, since each contains enzymes lacking in the other. Thus certain materials can only be formed by precursors passing through both sides of the unit. The foetal adrenal cortex is a specialization peculiar to primates and does not occur in other mammals.

Testosterone

Testosterone is the male sex hormone, secreted by the Leydig cells of the testis. There are three periods when its level in the blood rises. The first is in the foetus, where the rise begins at about 11 postmenstrual weeks and lasts probably till birth (see Chapter 4). During this time the testosterone first causes differentiation of the external genitalia to form a penis and scrotum, and then sustains the further growth of the penis. It may also have an action on the brain, differentiating the hypothalamus irreversibly to the male, non-cyclic, type (see pp. 56–7).

Blood testosterone falls rapidly in the first few days after birth

but a second period of high blood levels occurs during the subsequent 6 months, the peak being reached about 2 months after
birth. This rise has only recently been discovered and its function is
still unknown. By 6 months of age the level is down again to just
detectable amounts and it continues thus till puberty. Then a very
large increase occurs. At this time testosterone is responsible for
growth of the penis, prostate and seminal vesicles, the pubic, axillary and facial hair, the growth of muscles and the adolescent spurt
of the skeleton, especially the bones of the shoulders and vertebral
column. The presence of growth hormone is required for testosterone to exert its full effect on the muscles and on most of the
skeleton (see pp. 100–101 below). Since facial hair grows late and
pubic hair early in the sequence, either the hair follicles at the two sites
are unequally sensitive to differences in dose (pubic hair at low
dose, facial hair only at high dose) or else there is a sequential
maturation of testosterone receptors, this occurring earlier at the
pubic site than at the facial. The latter is the more likely possibility,
since when testosterone is given to a patient lacking it, pubic hair
grows at once, together with the penis and shoulders; but even at
high doses the beard responds only much later. Testosterone secretion is controlled by pituitary L H (luteinizing or Leydig-cell-
stimulating hormone).

Oestrogens
The ovaries secrete female sex hormones, collectively known as
oestrogens. The main one is oestradiol. Its blood level is low until
puberty, when a large increase occurs, causing growth of the
breasts, uterus and vagina, and development of the associated
vaginal glands. It also causes growth of parts of the pelvis. When
menstruation starts, oestradiol levels fluctuate regularly with the
phase of the cycle. Oestradiol secretion is controlled by pituitary
F S H (follicle-stimulating hormone).

Gonadotrophins
There are two pituitary gonadotrophins: follicle-stimulating hormone (F S H) and luteinizing or Leydig-cell-stimulating hormone
(L H). Both are secreted at low levels during childhood and increase
sharply at puberty.
    In males, L H is secreted, like G H, in a pulsatile fashion. In bulls
secretion follows the sight of a receptive cow or even a perfumed
model. It is uncertain how far similar considerations apply to man.

The first indications of the pubertal rise are pulses of LH released during sleep; subsequently LH is released during the daytime also, until by the end of puberty the total amounts released by day and by night are about equal.

LH in females is secreted in a cyclic fashion and interacts with FSH to control the menstrual cycle. The egg grows under the influence of FSH and is shed; the cells immediately surrounding it remain behind in the uterus and form a body called the corpus luteum: LH stimulates this body to produce the hormone progesterone. Progesterone maintains the uterus in a state receptive to the implantation and growth of the ovum should it be fertilized; if no fertilization and implantation occurs, LH levels decline, the uterus sheds its lining in menstruation, and the cycle starts again.

FSH in males causes growth of the seminiferous tubules, the sperm-producing parts of the testis. These occupy 90% of the volume of the testis, so the testicular enlargement at puberty is almost entirely due to them and not to the Leydig cells, however much the number of these increases. FSH is necessary for the growth of sperm, just as it is for the growth of the eggs. Much of the initial growth of eggs takes place during late foetal life, and in girls, though not in boys, FSH is especially high around 20–30 postmenstrual weeks.

Both gonadotrophins are released from the pituitary by LHRH, hypothalamic luteinizing-hormone–releasing hormone, which also stimulates their synthesis and storage. The feedback which shuts off LHRH is through oestradiol in the female and through testosterone and probably yet another testicular hormone, inhibin, in the male.

Prolactin

Prolactin, secreted by the pituitary, is necessary for the secretion of breast milk in the adult, but seems to play little or no part in the endocrinology of childhood. The blood level in boys stays the same from early childhood onwards, with no change at puberty. In girls, the pre-pubertal level is the same as in boys, but a small rise occurs in late puberty: perhaps this is responsible for the filling out of the breast beneath the protuberant areola which causes the passage from the oestrogen-induced stage 4 to the adult stage 5.

Insulin

Insulin is produced by the islets of Langerhans, collections of cells in the pancreas. It causes glucose to be absorbed into cells, and stored as glycogen in the liver and muscles; lack of it results in sugar diabetes. It is present in children just as in adults and has no particular action as far as growth is concerned except that it must be present in normal amounts for normal growth to occur. Diabetic children whose diabetes is well controlled by injected insulin and a suitable diet grow quite normally, but even a small degree of laxity in the control produces stunting and retardation of growth.

The Endocrinology of Puberty

The precise details of how all these hormones produce the changes of puberty are not fully known, though they have become much clearer in the last few years. Methods are now available for estimating the small amounts of hormones found in blood and urine, but the pulsatile nature of much hormone release makes single blood estimations in many cases useless. A small suction pump has been devised which, attached to a tiny tube inserted in the arm vein, sucks out blood continuously at the rate of only 1 ml per hour. The pump can be carried in a belt or shoulder sling and does not interfere with normal activities. The tube can be left in the vein for 24 hours or longer. Thus 'integrated 24-hour blood levels' of hormones can be obtained. Even this value, however, does not tell us all we need to know about actual rates of secretion, for some hormones are cleared out of the blood fast and others slowly, and the rates differ under different circumstances. Difficult though these technical problems are, they are still not as hard as the problem of studying a group of normal subjects longitudinally throughout puberty. Such a study requires repeated blood and urine samples as well as the usual body measurements. Very few such studies have been or are being made. Yet without them we shall remain in considerable ignorance about the evidently very complex hormonal events that underlie the changes of puberty.

 The adolescent growth spurt is produced by the joint actions of androgens and growth hormone. The 24-hour integrated blood levels of GH are higher in children than in adults, but there is no rise during the adolescent spurt. Nevertheless, the usual level of GH must be present for testosterone to produce its full growth-effect on the muscles and on the bones of the limbs and shoulders. In the absence of GH the height spurt is only about two-thirds of normal, and the shoulder-width spurt even less. Curiously enough the vertebral-column growth spurt in boys can apparently occur without

GH. In the absence of testosterone, of course, no spurt takes place at all. In girls the height spurt is also only about two-thirds of normal in the absence of growth hormone; and the oestrogen-induced growth of hip width is likewise reduced.

Initiation of Puberty

The first event in the sequence of puberty, immediately preceding the morphological changes, is an increased secretion of the two gonadotrophic hormones of the pituitary. This is caused by an increase in secretion of LHRH by the cells of the hypothalamus. At first, the LHRH produces a relatively small secretion of LH and FSH, either because the LHRH receptors in the pituitary cells need some stimulation to sensitize them or because the LH- and FSH-synthesizing cells require some practice. Gradually the response increases, the blood levels of LH and FSH rise, and the peripheral events follow.

see 98

The stimulus to the secretion of LHRH is as follows (in girls): during childhood a feedback system already exists, with LHRH stimulating the production of FSH, FSH stimulating production of oestrogen, and oestrogen inhibiting the production of LHRH. The system operates, however, at a low level, and the amounts of oestrogen circulating are insufficient to stimulate breast or uterus growth. Then something occurs to render the hypothalamic sensor of oestrogen less sensitive. The low level of oestrogen fails to stimulate it; LHRH is not inhibited and begins to rise. In consequence, FSH rises, causing oestrogen to rise to a level sufficient to bring about the pubertal changes. Only when this high level of oestrogen is reached is the sensor stimulated to turn off LHRH secretion. The feedback circuit is thus re-established, but at a higher level.

Whatever it is that turns down the sensitivity of the oestrogen sensor, it represents the culmination of a long chain of maturational events. One thinks of a series of clocks, each linked to the next, which have successively to run their courses before the strikers fall. Chronological age is a poor guide to the state of advancement of the clocks, as has been explained in the previous chapter. Bone age is a better guide, at least as far as menarche is concerned (see pp. 82–3). Indeed Drs W. A. Marshall and Y. de Limongi (1976) of the Institute of Child Health have shown how bone age may be used in conjunction with chronological age to predict when menarche will occur. Chronological age enters the prediction because the older the girl, the more likely is menarche to occur soon. Bone age aside, British girls less than 13·0 years old are most likely to have menarche at 13·0, but those more than 13·0 years old are most likely to have menarche tomorrow. This is a somewhat unusual situation in predictive statistical work and failure to recognize that this is so has led to invalid claims that age at menarche can be predicted from height and weight or that menarche takes place at a certain 'critical' value of weight, or weight for height (sometimes erroneously referred to as 'fatness'). Bone age is the only factor we know which does link more closely than chronological age with age at menarche, yet it itself is clearly only a symptom of some underlying maturity state, presumably located in the brain.

The nature and control of this maturity state are both obscure. One feels we should be able to specify a hormonal cause of early and late maturation, and isolate and synthesize a tempo-hormone. But we are far from any such achievement. We have to remember that even bone age fails to predict the age of initiation of breast development or the first enlargement of testes. Children differ in how closely linked all their pubertal events are; in some the 'clocks' for breast growth and for menstruation run far apart, in others they are closely synchronized. External events may slow down or stop the clock. The stimulus which changes sensitivity of the oestrogen sensor comes from the nervous system. Other nervous impulses, sometimes contingent on stress and emotion, may interfere with the smooth running of events. Thus R. H. Whitehouse and the writer have seen quite unusually 'fractured' curves of growth and pubertal development in girls translated to unfamiliar boarding schools at various times in puberty.

One last factor, indicated above, seems to be of importance specifically in female primates. The oestrogen sensor's responsiveness is not merely lessened; at a certain moment in late puberty it seems to be flipped right over. Suddenly a high oestrogen level turns on LHRH secretion instead of turning it off: the feedback changes from negative to positive. From this flows the complex physiology of the primate menstrual cycle. The causes of the sudden reversal in sensitivity are at present totally obscure; we do not even know whether they are hormonal, or nervous, or both.

In males the feedback circuits seem to be similar, with testosterone in place of oestrogen. No late flip-over of the sensor, however, occurs.

CHAPTER 8
Growth and Development of the Brain

Despite the overwhelming importance of the brain in the life of man we know all too little about its growth and about the development of its intricate organization. Anatomical studies of brain structure are immensely laborious and few workers have had the courage, persistence and technical support needed to carry out thorough morphological studies of the brains of children or even, indeed, of young animals. Though biochemical studies of brain composition are much easier they inevitably gloss over just that detail which is the chief feature of the brain's organization. The time is certainly coming when immunochemical methods will allow the same sort of fine distinction between individual cells that the microscope and the tiny probes of the electrophysiologist permit. But at present structural studies remain the solid base on which understanding of the brain rests. This means making three-dimensional reconstructions based on light- and electron-microscopy, aided by various computer methods. The elegance and beauty of the pictures thus produced, for example by Sidman and Rakic (1974) of Harvard University, make molecular biology look dull and molecular structure simple.

There is great difficulty in obtaining a sufficient number of brains to make a study of normal development from children dying of accidental causes, or at least of diseases which do not affect the brain. Foetuses obtained from abortions are relatively numerous and the early intra-uterine portion of the curve of brain growth is accurately known. But the important part of the curve in the last weeks before birth, and in the period from birth to 2 or 3 years, is still poorly mapped and its details are in some dispute.

Growth of the brain is reflected, albeit somewhat indirectly, in growth of the skull, and for this, at least, we have accurate post-natal data. We also have good prenatal longitudinal data from measurements by ultrasound (see Chapter 3, pp. 47–8) of maximum head breadth and, more recently, of head circumference. The head develops earlier than any other part of the foetal skeleton, and the peak velocity of head breadth is probably reached a little before 13 postmenstrual weeks, though a relatively high velocity continues till about 30 weeks. Velocity curves for length and breadth of the cerebrum also show maxima at about 13 weeks, which is to be expected since experimental work has shown that the growth of the vault of the skull depends on, and is controlled by, the growth of the brain below it.

Head circumference has its peak velocity at about 15 to 17 post-menstrual weeks, a little later than head breadth. This may be because the cerebellum, situated at the back of the skull (and hence reflected in head circumference though not in head breadth measurements) grows later than the cerebrum. The relatively high velocity continues till 32 to 34 postmenstrual weeks, but after this the deceleration is very rapid. Six months after birth, head circumference velocity is down to about 15% of its 34-week value, and by 1 year after birth, to 7%.

The velocity curve for the weight of the whole brain rises somewhat more slowly than the curve for head circumference, and reaches its peak around 32 postmenstrual weeks, according to the most numerous data, summarized by Dr Donald Cheek of Melbourne (see also Schultz *et al.*, 1962). Its deceleration thereafter is quite swift; by 1·0 years the velocity is reduced to about 25% of the peak value and by 2·0 years to less than 10% of the peak. The early development of the brain compared with that of most other organs means that from early foetal life its weight is nearer to its adult value than any other organ except perhaps the eye. At birth the brain averages about 25% of its adult weight; at 6 months nearly 50%; at 2 years about 75%; at 5 years 90%; and at 10 years 95%. This contrasts with the weight of the whole body, which at birth is about 5% of the young adult weight, and at 10 years about 50%.

However, total brain size or brain weight is not a very satisfactory measurement. Different parts of the brain grow at different rates and reach their maximum velocities at different times. There are maturity gradients in the brain no less than in the bones of the skeleton. Figure 35 shows the percentages of their volume at birth reached at various times of foetal life by the spinal cord, the pons and medulla (the hind part of the brain), the mid-brain, the cerebrum (which here includes corpus callosum, basal ganglia and diencephalon with the thalamus and hypothalamus) and cerebellum. The spinal cord, hind-brain and mid-brain are most advanced, and the cerebrum next (for a picture of the brain, see Figure 36). The cerebellum has its peak velocity considerably later than the cerebrum, an arrangement which seems to be general amongst mam-

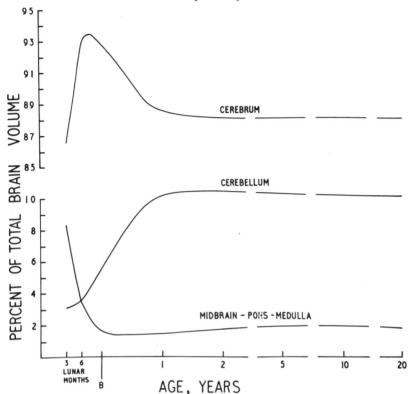

Fig. 35. Percentage of their volume at birth reached at earlier months by parts of the brain and spinal cord. Cerebrum includes hemispheres, corpus striatum and diencephalon. (From Tanner, 1977; data drawn from Dunn, 1921)

mals. (The cerebellum is the part of the brain concerned chiefly with the control of fine movements; it has a more uniform structure than the cerebrum and seems to many investigators a somewhat easier proposition to tackle.)

Even these curves are still far too non-specific. The brain is made up of two fundamentally different sorts of cells, the neurons proper (or nerve cells), and the neuroglia (or supporting cells). The neurons are the cells which transmit impulses. Each consists of a cell body, with a nucleus as in other cells, and a cytoplasm drawn out into a very large number of fine wire-like processes called dendrites. In most cells one such process is much larger and much longer than the others and is called the axon. In a motor nerve of the cerebral cortex, for example, the axon runs all the way down from the head to cells in the spinal cord. Axons themselves usually have many smaller branches. The dendrites are mostly short and form a pattern of arborization or branching, the intricacy of which invokes comparison with the statistics of the solar system. There are about 10^{12} or a million million neurons in the brain, and the average cerebral cortex neuron would seem to have some 30,000 nerve processes terminating on it, probably coming from one-tenth or so of that number of cells. The neurons are not tightly packed together like cells in most other tissues; there are narrow gaps between each of them, filled with tissue fluid. The connection with other cells is made through branches of the axon coming into close proximity with the dendrites of other cells, though without joining. Messages are passed across the tiny gaps by chemicals released at the nerve endings. It is thought that the most important clue to the functional capacity of a brain is the 'connectivity' of its neurons, i.e. the number of connections each cell makes with the others.

The neuroglia, which occupy about half the cellular volume of the brain, do not carry messages like the neurons. They are the support links, in the logistical sense, and act as intermediaries between neurons and the blood supply; a number have one process attached to a capillary blood vessel and the other intertwining amongst the surrounding neurons. They are smaller and more numerous than the neurons. They probably transmit glucose, amino-acids and other substances to neurons for the production of energy and the manufacture of structural protein and the chemical messengers. There is also evidence that during development some sorts of neuroglia act as scaffolding along which neurons move from their point of production to their final position. Other neuroglia manufacture the myelin sheaths which surround the axons. Thus neuroglia have clearly a crucial role in the nervous system.

Neurons and neuroglia develop at quite different times, at least as concerns cell division. Most of the neurons in the cerebrum are formed during the period from about 10 to 18 postmenstrual weeks; that is to say, their nuclei are formed then, surrounded by minimal cytoplasm. Axon and dendrite growth comes later; in some cases, it would seem, very much later indeed. Glial cells in the cerebrum begin to be formed around 15 postmenstrual weeks and continue being formed, though at a decreasing rate, until some time in postnatal life. It is said that new neuroglia cells are formed in the cerebrum up to about 2 years postnatally, but, surprisingly, in the cerebellum only up to about 15 months. However, these are generalizations based on chemical rather than structural evidence.

Cerebral Cortex Development

We owe much of our knowledge about the development of the structure of the cerebral cortex to the enormously painstaking studies of J. L. Conel, of Boston, who between 1939 and 1967 published analyses of the cortex at birth, 1 month, 3 months, 6 months, 15 months, 2 years, 4 years and 6 years. Additional studies of pre-term babies of 6, 7 and 8 months, and of children of 8 years, have been added by Conel's associate, T. Rabinowicz.

The cortex is identifiable at about 8 postmenstrual weeks, and by 26 weeks most of it shows the typical structure of six somewhat indeterminate layers of nerve cells (the grey matter) on the outside, with a layer of nerve fibres (the white matter, consisting of outgoing and incoming axons) on the inside. At first the cells are small, consisting mostly of nuclei with few and small processes. As they develop, axons and dendrites grow, protein appears in the cytoplasm, and many of the axons acquire an insulating sheath of myelin, the thicker fibres in general getting more, the thinner ones less.

From these cellular changes a series of criteria for maturation of parts of the cortex can be developed in just the same way as the criteria for skeletal maturity are developed from a consideration of the appearance of the bones. Conel used nine such criteria, amongst them the 'density' of neurons (i.e. the number of neurons per unit volume of cortex), which decreases as axons and dendrites grow in between them; size of neurons, which increases; nature of substances in the cytoplasm; length of axons and dendrites; and degree of myelination.

Two clear gradients of development occur during the first 2 years after birth. The first concerns the order in which general functional areas of the brain develop, the second the order in which bodily localizations advance within the areas. The most advanced part of the cortex is the primary motor area located in the pre-central gyrus (see Figure 36). This is the area whose cells initiate most movement. Next comes the primary sensory area in the post-central gyrus where nerve fibres mediating the sensation of touch end; then the primary visual area in the occipital lobe where nerve paths from the retina end; then the primary auditory area in the temporal lobe. All the association areas (where the primary impulses are compared

LATERAL VIEW

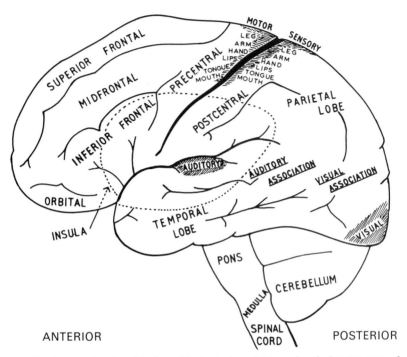

FIG. 36. Lateral and medial views of brain, to show divisions of cerebral cortex, areas of
localization of function, and the location of thalamus, cerebellum, pons, medulla and spinal
cord: (*a*) lateral view of cortex; (*b*) medial view of cortex and subcortical structures
(From Tanner, 1977; after various maps)

and integrated with other impulses) lag behind the corresponding
primary ones. Gradually the waves of development spread out, as it
were, from the primary areas. Thus in the frontal lobe the parts
immediately in front of the motor cortex develop next and the tip of
the lobe develops last. The gyri on the medial surface of the hemisphere and in the insula are generally last to develop; the hippocampal and cingulate gyri lag behind most areas of the frontal
lobe; and the insula develops later than any part of the frontal lobe
examined.

   Within the motor area, nerve cells controlling movements of the
arms and upper trunk develop ahead of those controlling the leg.
The same is true in the sensory area. This corresponds to the

MEDIAL VIEW

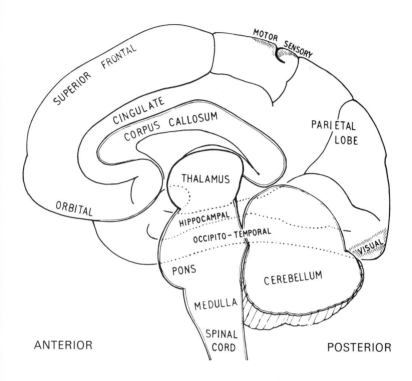

greater maturity of the arm relative to the leg in bodily development and also to the infant's capacity to control his arms more than
his legs. The leg areas remain the least developed up to 2 years and
presumably somewhat beyond. In the association areas we would
not expect gradients of this sort, since little or no localization by
bodily areas occurs there. Conel, in fact, describes development
outside the primary areas as proceeding in a fairly uniform
sequence within each lobe, so that the cortex at birth is little
developed and its appearance does not suggest that much cortical
function is possible. By 1 month the appearance of the cells of the
primary motor area of the upper limb and trunk suggests it may be
functioning. By 3 months all the primary areas are relatively

mature. At this age the motor area is the most advanced part of the cortex, and within it the hand, arm and upper trunk are the most advanced sections. The precise timings should be looked on as approximate, pending further research.

D. P. Purpura, also of Harvard University, found that the cells of the primary visual area of the cortex had their chief burst of maturation in the relatively brief period between 28 and 32 postmenstrual weeks. At this time the dendrites grew and 'spines' appeared on them, which is believed to signify full maturation; thus the visual cortices of the pre-term 8-month-old and the full-term neonate are said to be closely similar. He places development of the motor cortex also a little earlier than Conel, without disputing, however, the maturity gradients within the cortical areas.

To return to Conel's series; by 6 months most areas appear more mature than at 3 months and many of the exogenous fibres coming to the cortex have become myelinated, i.e. covered with an insulating sheath which in many nerves is essential for proper conduction of impulses. Between 6 and 15 months the rate of development is greatest in the temporal lobe and in the cingulate gyrus and insula; next greatest in the occipital lobe; and least in the parietal and frontal lobes, which have already passed through most of their development. The primary motor area is still slightly in advance of all the others, but within it the leg still remains behind the rest. The child's behaviour reflects this; he is beginning to control his hands and arms quite successfully, but he controls his legs less well. The visual association area has matured somewhat and is ahead of the auditory. By 2 years the primary sensory area has caught up with the primary motor, and the association areas have developed further. But some areas, most notably the hippocampal and cingulate gyri, particularly the insula, are still quite immature.

It is clear from the studies on myelination by P. I. Yakovlev and his colleagues that the brain goes on developing in the same sequential fashion at least till adolescence and perhaps into adult life. Myelination of nerve fibres is only one sign of maturity, and fibres can and perhaps sometimes do conduct impulses before they are myelinated. But where the two studies overlap the information from myelin studies agrees well with Conel's information on nerve cell appearances. As a rule the fibres carrying impulses to specific cortical areas myelinate at the same time as those carrying impulses away from these areas to the periphery; thus maturation occurs in arcs or functional units rather than in geographical areas.

A number of tracts have not completed their myelination even 3 or 4 years after birth. The fibres which link the cerebellum to the cerebral cortex and which are necessary to the fine control of voluntary movement begin to myelinate only after birth, and do not have their full complement of myelin till about age 4. The reticular formation, a part of the brain especially developed in primates and man and concerned with the maintenance of attention and consciousness, continues to myelinate at least until puberty, perhaps beyond. Myelination is similarly prolonged in parts of the forebrain near the mid-line. Yakovlev suggests that this is related to the protracted development of behavioural patterns concerned with metabolic and hormonal activities during reproductive life.

Throughout brain growth from early foetal life the appearance of function is closely related to maturation in structure. Fibres of the sound-receiving system (the 'acoustic analyser') begin to myelinate as early as the 6th foetal month, but they complete the process very gradually, continuing until the 4th year. In contrast, the fibres of the light-receiving system or optic analyser begin to myelinate only just before birth but then complete the process very rapidly. Yakovlev points out that in foetal life the sounds of the functioning of maternal viscera are the chief sensory stimuli. They are evidently not perceived at a cortical level; but at a subcortical one the analyser is working. After birth, however, visual stimuli rapidly come to predominate, for man is primarily a visual animal. These signals are very soon admitted to the cortex; the cortical end of the optic analyser myelinates in the first few months after birth. The cortical end of the acoustic analyser, on the other hand, myelinates slowly, in a tempo probably linked with the development of language.

There is clearly no reason to suppose that the link between maturation of structure and appearance of function suddenly ceases at age 6 or 10 or 13. On the contrary, there is every reason to think that the higher intellectual abilities also appear only when maturation of certain structures or cell assemblies, widespread in location throughout the cortex, is complete. Dendrites, even millions of them, occupy little space, and very considerable increases in connectivity could occur within the limits of a total brain weight increase of a few per cent. The stages of mental functioning described by Piaget and others have many of the characteristics of developing brain or body structures and the emergence of one stage after another is probably dependent on (i.e. limited by) progressive maturation and organization of the cortex.

Hemispheric Specialization

The two hemispheres of the human brain are not mirror images of each other. Some tasks are predominantly done by one side, some predominantly by the other. The brain itself is structurally asymmetrical to a far greater degree than the limbs or eyes or ears. The upper surface of the temporal lobe is on average considerably larger on the left side than on the right, a difference established during foetal life. This is the area which is chiefly concerned with the analysis of sounds, and indeed it is the left hemisphere which predominates in the activities of receiving, processing and producing, language. As Levy puts it 'The right hemisphere may know that "cat" means a furry small pet with claws, but it does not know that "cat" rhymes with "rat".' That association is made in the left hemisphere. In contrast, the right hemisphere plays the major part in the processing of spatial information, either visual or tactile. It notes visual similarities in preference to conceptual ones; it codes input in terms of images while the left hemisphere codes input in terms of linguistic descriptions. The relation of brain differentiation to handedness is complex and not properly understood at present, but some psychologists working in this field think that the brains of lefthanders are less clearly lateralized than those of righthanders, and most are agreed that severe problems of school abilities and behaviour may arise if differentiation of the hemispheres fails to take place normally. S. F. Witelson, for example, thinks that developmental dyslexia is due to having 'two right hemispheres and none left'.

There are several instances in other organs of the body where lateralization is dissimilar in the two sexes, and recently Witelson has found evidence of a sex difference in the timing at which lateralization of spatial information processing is established. Boys from at least age 6 (the lowest age tested) processed tactile spatial information better in the right hemisphere; girls processed it equally well in both sides at least till age 15 (the highest age tested). The average of the two sides was the same for both sexes. This work needs confirmation and extension, but may be of considerable importance in demonstrating a pre-pubertal sex difference in brain function, in the sense of one sex specializing earlier than the other. Boys have long been known to have a higher incidence than girls of reading and language disabilities, and reading, at least, involves spatial as well as linguistic processing. It would be quite in keeping with other aspects of sex differences if the male specialized away from the basic pattern, and ran some excess risk in doing so.

Influences on Brain Development

To what extent environmental stimulation can influence brain maturation or organization is not clear. Ramon y Cajal, the great Spanish pioneer of brain histology, believed simply that use of a cell caused axons and dendrites to grow, or, in modern terms, increased a cell's connectivity. As yet, there is no clear direct evidence that this is the case. However, in the last few years a body of knowledge has emerged which indicates, perhaps, the beginning of true understanding of how the nervous system generates its fantastically intricate and precise patterns.

Recently increased attention has been paid to the fine analysis of brain development in animals carrying genes causing brain abnormalities. As usual, the study of abnormality throws much light on normal processes; what happens in the absence of a certain class of cells, for instance, may give clues as to what that class of cells normally does (and this is the elementary level we are still at in studies of the brain).

The picture is one of much more movement and change within the central nervous system, or at least within the growing central nervous system, than we had previously imagined. The fact that no new nerve cells can be made after the first few months of life had somehow produced a picture of immutability wholly unlike our picture of, for example, liver cells or cells of the thyroid gland. Yet the dynamic state of the body constituents holds true as much for the brain as for any other part of the body. The turnover of amino-acids in even the adult brain is as high as that in the liver, and higher than that of any other tissue of the body. Since to supply its energy the brain only uses glucose, this means that the protein that makes up the structure of the cell is continuously being synthesized and degraded. (A small amount of amino-acid is made into transmitter substances but not nearly enough to account for the massive turnover.) The reason for the rapid turnover is unknown; the cells are usually thought of as unchanging year after year, but we cannot really be sure of this. The turnover may indicate that overt structural changes are occurring, of the sort that are in principle recognizable by the electron microscope. Or it may indicate continuous restructuring at a purely molecular level.

Much of the evidence has come from studies of the visual system. In all mammals very precise connections are made from the retina to the occipital part of the cerebral cortex. In those mammals such as the monkey and man, whose eyes are situated in the front rather than the side of the head so that the visual fields overlap and binocular vision is possible, half the nerve fibres from each retina cross to the opposite side in travelling to the cortex. This ensures that each tiny section of the cerebral cortex has fibres from both eyes terminating in it. Thus an object in space which is seen by both eyes generates impulses in each retina and these two sets of impulses end up exciting precisely contiguous sets of nerve cells in an area of the cortex. The visual cortex, indeed, consists of millions of columns or slabs of cells arranged with crystalline regularity, each column being a cellular machine equipped to analyse the visual events that occur in a tiny portion of each retina. The location of the event is given by the identity of the particular slab activated and the neurons in the slab analyse what type of event is occurring. The problem is to discover how this extraordinary regularity has been generated during growth.

An equal regularity is seen in a portion of the brain called the geniculate body, which is simply a relay station for nerves on their way from retina to cortex. The first set of fibres terminates there, and the impulses are taken up by a second set of neurons, which pass them on to the visual cortex. Now, in the rhesus monkey, the cells of the geniculate body are all generated within the space of a few days during early foetal life, and Rakic (1976) has shown that at first they are not separated into their distinct layers. The fibres from the left and right retinae arrive, and, to begin with, all is confusion, fibres from one side being diffusely intermingled with fibres from the other. Gradually the ordered chess-board pattern emerges, with clusters of fibres from one side alternating with clusters from the other. Experiments in other species have led to the belief that two sets of endings terminating on one neuron must somehow exchange information as to their precise place of origin. On this information depends the retention or loss of the endings. It is not too hard to imagine how this might happen in an animal subjected to visual experience during this time; if two terminations were often activated simultaneously because they represented corresponding parts of the left and right retinae, they would persist; if not, one would die or at least become inactivated. It is harder to understand how this could occur in the darkened world of the embryo. D. H. Hubel and T. N. Wiesel, at Harvard University, have shown that in the monkey the whole elaborate structure of the visual system is complete before birth. However, covering one eye in the visual cortex from birth onwards results in a diminution of the number of cell terminations subserving the covered side, and an enlargement of the clusters of terminations from the other side. If the eye is covered for a year no recovery is possible. It seems that nerve cells, like muscles, atrophy from disuse.

In the cat, much of the architecture is complete before the eyes are opened, but C. Blakemore and others at Cambridge University have demonstrated that the cell connections required for precise binocularity or stereopsis of vision need the stimulus of visual experience. They also have shown that a kitten brought up in a world consisting exclusively of vertical stripes develops many more cortical cells which analyse this dimension of shape than cells which analyse the horizontal dimension (whereas normally each are present equally).

Furthermore, it seems that during development some axons form transient connections with a whole series of cells before establishing their stable adult contacts. In toads, for example, the eyes move their positions relative to the head during growth, and in order to keep the images seen by the two eyes as they grow projecting on to the same area of the brain, a series of cell contacts are successively made and broken. In addition there is evidence in adult mammals that axons entering the spinal cord from the periphery make a much wider series of connections in the spinal cord than are illustrated in the anatomical textbooks. These terminations seem normally to be suppressed; they remain, in Merrill and Wall's phrase, as 'ghosts of the cell's childhood'. The conditions under which they may be activated again are at present not clear.

R. Mark, of the Australian National University, whose book *Memory and Nerve Cell Connections* is a masterly exposition of the subject, sums up the new view thus: 'The principle of supplying, by growth, an overabundance of nerve cells or synaptic connections and fixing the final patterns by discarding many of them seems to show up often in the development of the nervous system, especially in the fine details of connection formation.' Dendrites seem no longer such static entities; perhaps even in the adult they may be built and broken.

As for pathological influences, particularly careful studies of the effects of various agents on the rat's cerebellum have been made by

Altman and his colleagues in Indiana and by Balazs in London (see Patel *et al.*, 1975). The rat is born so much earlier in its development than man that the cell multiplication phase of most of the cerebellar neurons as well as that of the cerebellar glia takes place after birth. Experimental manipulations are therefore relatively easy to make. A very large dose of radiation delivered daily to the cerebellum from birth to 10 days postpartum practically stopped cell multiplication, and the weight of the cerebellum at the end of the treatment was reduced to 47% of normal. However, catch-up then occurred, both in weight and in the number of cells formed. The catch-up was not complete; weight and cell number were both about 60% of normal at 23 days. It seemed that the duration for which cell division could continue was not extended; the cell division clock was adhered to, and the animal ran out of time in which to make cells. This is a result consistent with studies on the effect of undernutrition in rats (p. 133 below). However, in another experiment, Balazs himself found the cell division period of the cerebellum to be stretched out somewhat by undernutrition, so the matter remains in doubt. The radiation, unsurprisingly, had little effect on cells already present when it started; they continued to increase in size quite normally.

CHAPTER 9

The Interaction of Heredity and Environment in the Control of Growth

Growth is a product of the continuous and complex interaction of heredity and environment. Modern biology has no use for simplistic notions, let alone the obscurantist slogans of the deliberately ignorant. Statements such as 'Height is an inherited characteristic' or 'Intelligence is the product of social forces' (or vice versa of course) are intellectual rubbish, to be consigned to the trash-can of propaganda. As explained in Chapter 2, what is inherited is DNA. Everything else is developed. We may write $G * E \xrightarrow[\text{time}]{} (GE)$ where G stands for genetic and E for environmental factors. The product (GE) develops through time, and the way G and E are related is shown by a star, not a plus sign nor even a multiplication. We say that the interaction of genetics and environment is non-linear, meaning that the effects do not in general sum (as would be indicated by a plus sign).

Suppose, for example, we have two children, of different genotypes (i.e. arrays of genes). Under the environmental circumstances represented, let us say, by a contemporary well-off Swedish home, A grows to be 5 cm. taller than B. Transplant the same two genotypes to a peasant environment in southern India, growing fitfully in the face of recurrent famines, chronic infections, physical labour and heat. Both children will probably end up smaller. But what non-linear interaction tells us is that B is unlikely to be still 5 cm. smaller than A. He may be the same height or even taller in this different environment, for his genes may be more suitable for the regulation of growth in marginal circumstances.

Some interactions are more critical and dramatic. When a grey-lag goose hatches from its egg it has an inborn reaction to follow the first large moving object it sees. Normally this is the mother. But, as everyone knows, it will follow Dr Konrad Lorenz just as well if he should creep by first, and it will for ever after treat him as mother. The distinction between environment and heredity is a man-made one and evolution makes or ignores it according to the laws of Darwin rather than Aristotle. Animals are born into expected environments; some occupy niches of exquisite precision and negligible width; others, like man, are relatively polyvalent and can interact, with varying success, with a wide variety of environments. The classic work of Hubel and Wiesel and their successors (see p. 115) provides a mammalian example of an essential interaction of genotype and environment. If one of an adult cat's eyes is covered for a few weeks and then the cover is removed the cat is a bit clumsy for some minutes, then quite normal. But if one of a kitten's eyes is similarly covered for as little as a week from the time it normally begins to use the eyes, then the kitten never develops proper binocular vision; the cells of the cerebral cortex concerned with vision by the covered eye fail to develop their full characteristics. The genes have brought the cells to the point in time where light to both eyes is expected. Then the interaction must occur, or disaster follows. The experience of binocular seeing is necessary during the critical or sensitive period.

The facts of interaction lead to a principle from which the reader may deduce, if he wishes, a whole social philosophy. Everyone has a different genotype. Therefore, for optimal development (optimal, that is, for the individual and his environment considered individualistically), everyone should have a different environment. This is of course impossible; but one may seek either to optimize the interaction or to minimize it. In relation to cattle and other domestic animals we minimize the variation in both genotype and environment, aiming at a uniform product of maximal usefulness to ourselves. Such a strategy is foreign to successful evolution, and is fraught with the greatest danger. Should ecological circumstances change there is no reserve of genotypes suited to the new circumstances, no way forward and no way back. The true patrimony of a species is its genotypic variation.

So much for general principles. As we have seen (p. 28), it is DNA, not stature, nor blue eyes nor even blood group A that is inherited. The sequence in the cell goes DNA⟶RNA⟶amino-acid assembly⟶protein. The protein escapes from the cell to influence another cell in building a tissue or organ; the organ interacts with others in the embryo and foetus; the foetus interacts with the uterine environment; the child interacts with the complex and changing environment of the adult-created and self-created world. If we talk simply about cellular proteins, the steps from genotype to character are small and well protected. In everyday language one may certainly say that the blood group A protein is inherited. But beyond this the going is not so simple. It is a long way from the possession of certain genes to the development of a height of 2 metres, say. For such a character the only valid statement is thus: 'x per cent of the variation in the height of young adults brought up in the circumstances of comfortably-off urban homes in the temperate climate of central Holland in the 1960s is due to genotypic variation.' For similar young adults brought up in Saharan African villages x would be less, for the environment would intervene more decisively.

Figure 37 shows a pair of identical, monozygotic twins, studied through the courtesy of Mr James Shields of the Department of Psychiatric Genetics of the University of London Institute of Psychiatry. The twins were separated at birth. One was brought up in a normal manner in a sympathetic home, while the other was watched over by a neurotic and cruel relative who shut him for long periods in the dark and required permission to be sought (frequently it was refused) even for a drink of water. The two were brought together because of a search by Mr Shields for separated twins. Naturally, the deprived twin was more nervous than the other, but here we are more concerned with physical size, shape and body composition. The difference in size was considerable, amounting to 8·3 cm. (or about 3 ins.) in height. In skeletal shape, however, the difference was negligible; subtract the fat of the healthy twin and enlarge the photograph of the deprived, and the monozygosity is clear to see. It must be said that this is an extreme and highly unusual example of a size difference in monozygotic twins, but it shows dramatically the effect of continuous childhood undernourishment and neglect better, perhaps, than would a large series of individuals contrasted with unrelated controls. (The only caveat is that we do not know the birthweights; part of the persisting size difference may derive from differences in intra-uterine growth.

In the remainder of the chapter we shall necessarily leave this holistic approach and discuss in order:

1. Genetics of size, shape and tempo of growth
2. Effects of nutrition on growth and tempo of growth
3. Differences between races
4. Climatic and seasonal effects on growth
5. Effects of diseases
6. Psychosocial stress
7. Effects of urbanization
8. Effects of socioeconomic status and numbers in the family
9. Secular trend

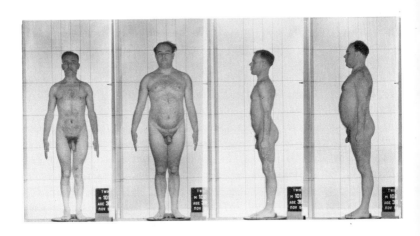

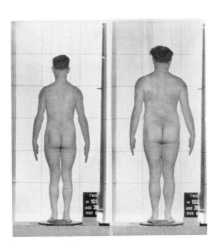

Fig. 37. Monozygotic twins brought up apart and under widely different circumstances from birth (see text)
(Twins seen by courtesy of Mr James Shields, 1962)

1. Genetics of Size, Shape and Tempo of Growth

Factors affecting the rate or tempo of growth must be considered separately from factors affecting the size, shape and body composition of the child. The genetical control of tempo seems to be independent of the genetical control of final adult size, and, to a large extent, of final shape. Equally, environmentally-produced changes in tempo do not necessarily affect final size or shape. Indeed, size and shape themselves seem to be separately controlled, both by genetical and by environmental factors. The genetical control of shape is much more rigorous than that of size, presumably because shape represents chiefly how the cells are distributed, while size represents more the sum of sizes of the various cells. As explained in Chapter 2, the number of cells is fixed early, in the relative security of the uterus; their size continues to alter during much of childhood, and in some instances, such as the fat cell, throughout life.

TABLE 1

	Mean difference in length (cm.)		Correlation coefficient	
	MZ *pairs*	DZ *pairs*	MZ *pairs*	DZ *pairs*
Birth	1·8	1·6	0·58	0·82
3 months	1·4	1·6	0·75	0·72
6 months	1·3	1·9	0·78	0·65
1 year	1·3	1·8	0·85	0·69
2 years	1·1	2·4	0·89	0·58
3 years	1·1	2·9	0·92	0·55
4 years	1·1	3·2	0·94	0·60

Mean differences between lengths of monozygotic twin pairs (∼ 140 pairs) and same-sexed dizygotic twin pairs (∼ 90 pairs) from birth to 4 years, and within-pair correlation coefficients
(From Wilson, 1976)

Monozygotic twins, who have the same genotype, usually resemble each other very closely indeed when brought up under similar circumstances. Table 1 shows the average differences in length or height from birth to 4 years between pairs of monozygotic and of dizygotic same-sexed twins, from the magnificent Louisville data of Wilson. Dizygotic same-sexed twins resemble each other genetically no more closely than any other brothers or sisters, since they arise each from a different fertilized ovum. At birth the monozygotic pairs were actually less alike than the dizygotic, but this situation rapidly changed. The increasingly close similarity of the values in monozygotic twins reflects size, shape and tempo combined. The differences between pairs at birth are probably partly due to asymmetrical division of the original ovum, one twin getting just a little more cytoplasm than the other, and partly to their different positions in an overcrowded uterus. The twin who is smaller at birth, by however little, usually remains smaller throughout life. When there is a large difference in size at birth, indicating some real intrauterine distress to the smaller twin, then the differences later are quite large. Babson and Phillips followed up 9 pairs of monozygotic twins, born in hospital in Portland, Oregon, with the unusually large difference of over 25% in weight. The smaller twins averaged 5·6, 6·0 and 6·8 cm. shorter at 8, 13 and 18 years of age, and were correspondingly lighter. In all cases the smaller twins scored lower, also, on tests of mental ability.

The degree to which height is controlled by genotype when environmental circumstances are adequate is reflected in the variation within families compared with the variation amongst a population. The range of variation in adult height, represented by ± 2 standard deviations around the mean, is about 25 cm. for most male populations, 16 cm. amongst brothers and 1·6 cm. amongst monozygotic twins brought up together. Height is generally said to be controlled by many genes, each of small effect. Such small-effect genes are called 'polygenes' and are believed to be located in parts of the chromosomes distinct from those carrying 'major' genes. The notion that height was controlled by polygenes arose because of the continuous and Gaussian distribution of height in the population (see Chapter 1, pp. 19–21). It is true that the summation of many small effects does indeed bring about exactly this type of continuous distribution; but it is also true that a distribution virtually indistinguishable from it may be brought out by as few as five or six major genes interacting. The question is thus an open one.

Little is known about the genetics of shape. Some body measurements show higher correlations between parents and grown-up children than others. In a recent study of 125 Belgian families, containing 282 grown-up children, C. Susanne found the correlations partly displayed in Table 2. In height each parent contributes equally to each offspring; despite all popular belief to the contrary, there is no tendency for daughters to resemble their mothers and sons their fathers: the four correlation coefficients (0·52, 0·47, 0·52,

TABLE 2

Correlations between measurements of parents and grown-up children: 125 Belgian families

Measurement	Parent–child	Brother–brother	Sister–sister	Father–son	Mother–daughter	Father–daughter	Mother–son
Height	0·51	0·53	0·57	0·54	0·47	0·52	0·53
Arm length	0·49	0·37	0·51	0·47	0·57	0·53	0·39
Sitting height	0·37	0·21	0·35	0·41	0·39	0·29	0·38
Biacromial diameter	0·33	0·42	0·46	0·09	0·50	0·33	0·41
Bi-iliac diameter	0·49	0·49	0·45	0·51	0·53	0·43	0·49
Head breadth	0·35	0·37	0·32	0·42	0·33	0·41	0·22
Head length	0·28	0·36	0·44	0·18	0·40	0·17	0·37
Nose length	0·31	0·00	0·44	0·32	0·26	0·34	0·35
Interpupillary breadth	0·38	0·34	0·40	0·32	0·40	0·42	0·38
Ear height	0·31	0·26	0·43	0·24	0·33	0·32	0·39

(From Susanne, 1975)

0·53) do not differ significantly. Other data fully bear out this generalization. Farther down the table, the correlations for bi-iliac diameter (hip width) closely resemble those for height. But the biacromial diameter (shoulder width) values are less similar. The father–son correlation is very low, and the father–daughter is lower than mother–daughter. Amongst the head measurements, head breadth resembles height in pattern though all the correlations are rather lower. But in head length the father's correlation seems especially low, and in nose breadth brothers resemble one another not at all.

Interpretation of these correlations is not entirely straight-forward. Measurements much affected by environment have low correlations; relatives simply do not resemble one another very much. If a measurement is chiefly controlled by a few genes of which one or more shows dominance (i.e., a single dose gives the same effect as a double dose), the correlations decrease. The presence of sex-linked genes (i.e. genes on the x chromosome) causes the sister–sister correlation to exceed the brother–brother correlation; the father–daughter to exceed the father–son; and the mother–son to exceed the mother–daughter. No example of this pattern appears in the table. What is clear is that different measurements show different degrees and patterns of familial resemblance. These differences cannot be understood until the development of each measurement is studied and the physiological factors controlling its growth are clarified.

Not all genes are active at birth. Some are not switched on till later, and the products of others can express themselves only in the physiological surroundings provided by the later years of growth. Some genes only produce their effect in one sex, usually because the gene product needs the co-operation of either one or the other hormonal environment in order to exert its action. These genes are called 'sex-limited'. The effect may not be all-or-none; the type of central baldness that is quite common in men depends on a gene which in single dose causes an effect in males only. In a double dose, which rarely occurs, it produces a similar effect in women.

Sex-limited genes are quite distinct from genes that are sex-linked. The sex-linked genes affect alternately males and females in successive generations (so-called 'criss-cross inheritance'). Quite often in sex-linkage the affected males are worse than the affected females. This is because the females have two x chromosomes and therefore two of the genes in question, and one of them will be normal. Males have only one x chromosome and thus lack the single normal gene.

The inability of some genes to express themselves till after birth probably accounts for the fact that the resemblance in size between children and their parents is quite small during the first 12 to 18 months. The correlation between parents' heights and children's heights (or lengths) is only about 0·2 when the children are new-born but swiftly rises till it reaches its adult value of about 0·5 when the children are aged about 18 months. This means that during the first 18 months many babies change their centile positions for height and weight. Children with genes making for large size but born to small mothers move upwards through the centiles, and children born large but with genes making for small size move downwards. Professor David Smith and his collaborators (1976) give practical examples of this movement in babies in a well-nourished American middle-class population; the upward shift was mostly complete within 6 months after birth, while the downward shift took up to 18 months to accomplish. From age 2·0 till adolescence the parent–child correlation can be used to make standards for children's heights which allow for the heights of their parents. These standards, described in detail in Chapter 11, provide a more accurate means of diagnosing short stature than the ordinary population standards.

The correlation coefficients between height measurements of a child at successive ages and his own measurement as a fully-grown adult describe the curve shown in Figure 38. The correlation of length at birth with adult height is only about 0·3 since size at birth reflects uterine conditions much more than foetal genotype. Thereafter the correlation rises steeply and by age 3 is of the order of 0·8. This means that the adult height can be predicted from height at age 3, with an error which may amount to as much as 8 cm. either way. (It takes a very much higher correlation than 0·8 to guarantee good prediction in the individual case.) At puberty, the correlations diminish because of some children maturing early and some late, but if bone age is taken into account the prediction is restored. Tables of prediction of adult height from childhood height and bone age are discussed in Chapter 11.

The correlation coefficients between height of the parents and height of the child at successive ages describe curves very similar to those shown in Figure 38.

Tempo of Growth. The genetical control of tempo of growth (as

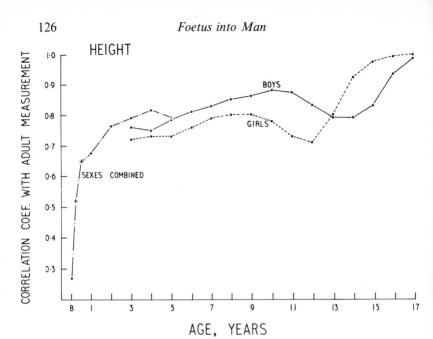

Fɪɢ. 38. Correlations between adult height and heights of same individuals as children. Sexes-combined lines (0–5) from 124 individuals of Aberdeen study with + points from Berkeley Growth Study, California. Boys' and girls' lines (3–17) from 66 boys and 70 girls of California Guidance Study. All data pure longitudinal
(From Tanner, 1962, p. 88, where sources of data are detailed)

contrasted to shape and size) is best shown by the inheritance of age at menarche. Monozygotic twin sisters growing up together under West European conditions reach menarche on average 2 months apart whereas dizygotic twins differ on average by 12 months. The sister–sister and mother–daughter correlations are about 0·4. Thus a large proportion of the variability in age at menarche under these conditions is due to genetical influence. It is thought that mother and father exert an equal influence on tempo of growth, so that a late-maturing girl is as likely to have a late-maturing father as she is to have a late-maturing mother.

This control of tempo operates throughout the whole process of growth, for skeletal maturity at all ages shows the same type of family correlations as menarche. The age of eruption of the teeth is similarly controlled.

Under reasonable environmental conditions the genetical control extends down to many of the details of the growth curves. This is demonstrated by the records of three sisters shown in Figure 39. In

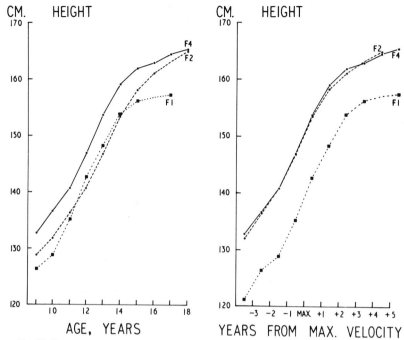

Fɪɢ. 39. Growth in height of three sisters: (*left*) height plotted against chronological age; (*right*) height plotted against years before or after year of peak height velocity. Note the coincidence of curves of ꜰ2 and ꜰ4 when equated for developmental age.
(From Tanner, 1962, drawn from data of Ford, 1958)

the left-hand figure the heights are plotted against chronological age and in the right-hand figure against developmental age, as given by years before and years after peak height velocity at adolescence. Two of the sisters have curves which are practically superimposable except that they are on a different time-base, one being almost a year in advance of the other. These two therefore differ radically in one parameter of their growth curve, but little in their other parameters. The third sister differs little from the other two in velocity (when plotted against developmental age the curves are 'parallel'), but markedly in absolute height.

2. Effects of Nutrition on Growth and Tempo of Growth
Malnutrition delays growth, as has been repeatedly shown by the effects of famine associated with war. Figure 40 shows the heights of children in Stuttgart, Germany, plotted at each year of age from

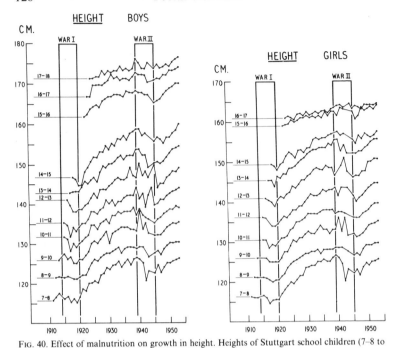

Fig. 40. Effect of malnutrition on growth in height. Heights of Stuttgart school children (7–8 to 14–15 Volkschule; 15–16 up, Oberschule) from 1911 to 1953: (*a*) boys; (*b*) girls. Lines connect points for children of same age and express secular trend and effect of war conditions. (From Tanner, 1962, drawn from data of Howe and Schiller, 1952, and personal communication)

1911 to 1953. There is a uniform increase at all ages from 1920 to 1940 (see discussion of secular trend, pp. 150–3 below) but in both world wars the height dropped as the food intakes of the children became restricted.

Children subjected to an episode of acute starvation recover more or less completely by virtue of their regulative powers, provided the adverse conditions are not too severe and do not last too long. Chronic malnutrition is another matter. Most members of some populations, and some members of all populations, grow to be smaller adults than they should because of chronic undernourishment during most or all of their childhood.

We should distinguish nutritional effects on tempo of growth, on final size, on shape and on tissue composition. Tempo seems usually to be the first thing affected; the undernourished child slows down and waits for better times. All young animals have the capacity to do this; in a world where nutrition is never assured, any species unable to regulate its growth in this way would long since have been eliminated. Man did not evolve in the supermarket society of today, but in small tribal communities, most of the period nomadic, following a usually precarious food supply. (Hence we can cope with periodic malnutrition better, perhaps, than with overfeeding.)

Adult size is affected by a less severe level of undernutrition than adult shape. Indeed, undernutrition in man does not alter shape significantly; a malnourished European child by no means acquires the short legs of the Asiatic. There is some slight evidence that the secular trend towards larger size includes a faint tendency towards increasing linearity of build, but the change is a very minor one.

In many populations the period when the child is most at risk from malnutrition, often combined with infection, is birth to 5 years. Though in some developing countries weights at birth are already low, in many (especially African countries) it is only after the first 6 months that weight-gains diminish. This often coincides with weaning and the substitution of high-starch, low-protein foods. It is also the age at which the mother's lactating ability declines so that satisfactory growth cannot be achieved by breast milk alone.

Professor John Waterlow of the London School of Hygiene and Tropical Medicine, and his colleagues there and in Jamaica, have recently made estimates of the energy requirements of infancy. Energy is measured in joules; the energy used simply for bodily maintenance in a child aged 1·0 years averages 330 kilojoules per kilogramme of body weight per day. The energy required for normal growth at this age averages about 20 kJ/kg/day. Normal infants use up some 80 kJ/kg/day in physical activity. Thus the maintenance requirement is a surprisingly high percentage of the whole. When calorie intake falls below 330 kJ/kg/day, as it commonly does in children in developing countries, then growth ceases. Even before this the energy margin for physical activity is eroded, and in the infant and child the restriction on exploration, play and social interaction that follows may be a more potent cause of delay in intellectual and emotional development than any nutritional effect on the nervous system.

Protein intake seems less important than was once thought. A child aged 1·0 years needs about 1·3 g milk protein per kg body weight; of this seven-eighths is used for maintenance and one-eighth for growth. If intake falls below this, growth ceases in 1 to 2 days. In the first 6 months, when the growth rate is higher, about three-eighths is used for growth. The protein–energy ratio in breast milk is about 7·5% in the first few weeks, falling to 5% at 2 to 3 months. This seems to be about the ratio required for healthy growth in childhood, and it is supplied by most foods, including cereals.

Figure 41 shows Dr Margaret Janes's work on the growth in height of two groups of boys in Ibadan, Nigeria. One group was drawn from the professional, highly-educated classes, living in considerable affluence; the second from indigent slum-dwellers in the

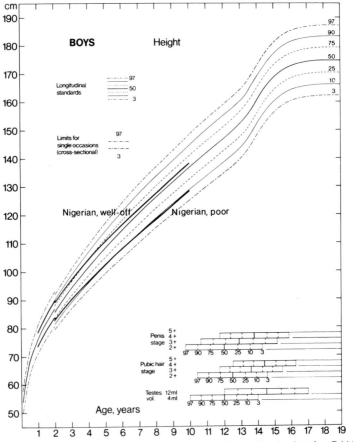

FIG. 41. Average growth in height of two groups of boys in Ibadan, Nigeria, plotted on British Standards: (*a*) well-off group; (*b*) indigent group
(Data from Janes, 1975, as quoted in Eveleth and Tanner, 1976)

market area of the town. The curves have been drawn on British standard charts. The well-off grow very similarly to British boys, while the 50th centile for the poor is no higher than the 10th centile of the British. There is no evidence that the two groups differ much in genotype, though they differ in other things besides nutrition. However, the major cause of the great difference in growth curves is certainly nutritional level.

The question whether undernutrition during foetal life or during the first 1 or 2 years after birth necessarily leads to a deficit in size or in mental function in adults has been frequently and sometimes heatedly discussed. It seems that children with severe protein–calorie malnutrition in early infancy due to malformations or malfunction of the gut (pyloric stenosis, cystic fibrosis) make a complete recovery in height after surgical correction, when brought up in reasonably well-off homes in a developed country. In the majority of studies, they have also caught up completely in intellectual ability, when assessed at ages 7 to 10 years. In one study, however, it seemed that starvation in the first 3 months after birth might have been responsible for a lowered score in auditory memory tests, though not in tests of general intelligence.

Evidently much depends on the circumstances after the severe episode of malnutrition is over. Children under 5 years old admitted to hospital in tropical countries with severe malnutrition (kwashiorkor or marasmus) have been followed up after leaving hospital. In most such children complete equality of height and weight with sib controls has been attained before puberty. In trying to isolate the effects of malnutrition from general social effects, sibs are more satisfactory controls than the general population, since they suffer some at least of the same environmental circumstances. Naturally, the children admitted to hospital come from poorly-off families where even the sibs are growing relatively badly.

Dr Steven Richardson of the Albert Einstein College of Medicine, New York, sums this work up very precisely in two generalizations: 'Where the histories of both the severely malnourished child and the comparison child suggest a similar level of nutrition over their life histories except for the presence or absence of an acute episode requiring treatment in hospital, and where both sets of children have experienced similar social, physical and biological environments, the children with an acute episode of malnutrition will not be smaller in somatic growth [at school age] than their comparisons' and 'Where the children with an early acute episode of malnutrition are more disadvantaged than their comparisons in the overall histories of malnutrition, social, physical and biological environments, then the malnourished children at school age will be smaller in somatic growth than their comparisons.' He adds, however, that his own work suggests that the second generalization needs qualification: 'Even when the children with acute malnutrition come from environments and backgrounds that are somewhat disadvantaged, complete catch-up may still occur' (Richardson, 1975).

A very well-controlled study of the long-term effects of undernutrition during foetal life was made by Z. Stein and M. Susser and their colleagues at Columbia University, New York. Between mid-October 1944 and May 1945 the central part of Holland, including the towns of Rotterdam, Amsterdam and Leiden, was subjected to a severe war-time famine. The official daily ration was 1,500 calories and the actual amount eaten only slightly more than this. Examination of the birth records in these three towns by monthly cohorts of birth showed that birthweight diminished by 9% in the cohorts who were exposed to the undernutrition in the third trimester of pregnancy (i.e. from 6 to 9 months). Babies exposed only in the first or second trimesters showed no weight reduction. Birth length was reduced by 2·5% and head circumference by 2·7% in the third trimester cohort. Maternal weight was 10% less than the control values taken after May 1945.

At age 19 the males of this group of children entered military service. At entry their height was measured and several tests of mental ability were administered. In addition the incidence of mild and severe mental retardation was ascertained for the relevant cohorts. Stein and Susser found that young men who had been undernourished in foetal life were no different in height, weight or mental ability from those who had not been undernourished. Nor was the incidence of mild or severe mental retardation any greater. They searched for differential effects of the famine exposure according to social class, family size and birth order, but found none. Their final conclusion is an impeccable summary of the present position: 'We believe we must accept that poor *prenatal* nutrition cannot be considered a factor in the social distribution of mental competence among surviving adults in industrial societies. This is not to exclude it as a possible factor in combination with poor *postnatal* nutrition, especially in pre-industrial societies' (*Famine and Human Development*, p. 236, their italics).

Experimental Models. A considerable amount of experimental work on animals has been done, designed to throw light on the effects of long-term undernutrition. As a result many and sometimes strident claims to certain knowledge have been made, especially in relation to the highly important and emotive subject of brain growth. Such claims should be treated with the utmost caution; our knowledge in this area at the present time is not at all great.

An early difficulty was the *naïveté* of physiologists about species differences. Much of the early experimental work was done on the rat, a species whose growth is about as different from man's as it is possible to find amongst the whole range of mammals. Rats are born at a far earlier stage in development than primates, at an age roughly corresponding to 16 postmenstrual weeks in man. Thus, undernutrition that is immediately postnatal in the rat corresponds, if to anything in man, to undernutrition of the foetus in mid-pregnancy. This does not result from maternal undernutrition (unless at catastrophic levels) but only from pathology of the placenta, a wholly different situation. Besides this, the rat bears litters not singleton foetuses, its whole endocrine system develops in a very different way from that of man, and there are enormous differences in behaviour and brain function.

The rhesus monkey and the New World cebus are much more suitable models. Dr Donald Cheek of Melbourne, Dr Don Hill of Toronto and Dr George Kerr and his associates of the Harvard School of Public Health have in particular studied the former, and Dr Hegsted and his co-workers also at the Harvard School of Public Health, have recently investigated the latter (Fleagle *et al.*, 1975). Primates all have the same characteristic growth-curve, although monkeys are born at a more developmentally advanced stage than apes, and apes at a more developmentally advanced stage than man. Endocrine development in monkeys is not dissimilar from man's and even the brain shares many of man's preoccupations.

Cheek and Hill studied malnutrition during the foetal period. The rhesus monkey (*Macaca mulatta*) has a pregnancy lasting 165 days from fertilization. Hill tied off some of the placental vessels at 100 days to cause relative placental failure and intra-uterine malnourishment in the foetuses. At birth the animals were severely underweight, averaging about 2·5 standard deviations below the mean of controls.

Cell numbers and cell size in the cerebrum were entirely un-affected; in the cerebellum a small reduction in the numbers of cells occurred. In this species the maximum velocity of cerebrum growth takes place from 80 to 100 days after fertilization, i.e. before the experimental intervention. Thus the negative result is quite ex-pected. In relating this to man, however, one should remember that the rhesus at birth has a brain weight 60% of the adult value whereas man's brain develops more slowly and at birth is only about 25% of the adult value. From the cerebrum-time-table point of view, the placental insufficiency studied by Hill corresponds to placental insufficiency near to birth in man (judging by the timing of maximum growth rates of brain weight curves). The cerebellum develops later than the cerebrum in both species (see Chapter 8), which probably explains the slightly greater effect of the malnutri-tion upon it.

Muscle bulk in the animals was reduced to 67% of that of con-trols. Most of this reduction was due to a smaller average cell size, with only a questionable effect on cell number. The liver, however, presented quite a different picture. Its weight was reduced to 75% of that of controls and the reduction was primarily in number, not size, of cells. This may relate to the fact that the liver is one of the few organs in mammals that has a capacity for regeneration (see Chapter 2, pp. 24–5). It is possible, though not established, that some catch-up in cell numbers could occur in the liver under circumstances in which this would be impossible in other organs.

The work on cebus has been in the postnatal period. The monkeys were separated from their mothers at birth and reared in a primate nursery with other monkeys. For the first 8 weeks they were fed a commercial human baby food and thereafter a synthetic liquid formula of which the control monkeys had as much as they wished. Undernutrition began at this time in two experimental groups, one simply fed 67% of the controls' diet and the other fed a very low-protein diet (2·8% of calories as protein) but in as large a quantity as desired. The three groups were followed for 20 weeks, then the two experimental groups were rehabilitated, eating as much of the control diet as they wanted. By the end of the 20-week experiment period the controls had doubled their weight, the calorie-deficient had increased their weight by 20% and the protein-deficient had scarcely gained weight at all.

Extensive physical measurements were made on these monkeys so that changes of shape as well as size could be followed. The growth of all parts of the skeleton slowed down, and did so to a greater extent in the protein-deficient than in the calorie-restricted group. Some growth occurred in nearly all skeletal measurements, however, even in the worse-off group. The shape changes could be important, theoretically for throwing light on the mechanism of growth, and practically as an aid in diagnosing undernutrition in human infants. J. G. Fleagle, D. M. Hegsted and their colleagues found that some skeletal relationships stayed quite constant under these conditions; e.g. the relative lengths of upper and lower arm. But feet and hands suffered more than legs and arms, and sitting height more than lengths of arms and legs. It seems that those dimensions which were farthest along their road to maturity were held up most (hands are ahead of forearms, feet ahead of calves, sitting height ahead of limb lengths, in cebus as in man; see Chapter 10, p. 164). 'In response to malnutrition', write the authors, 'the available resources are expended disproportionately in favour of parts with the greatest distance to cover in the future ... in long-term malnutrition such a distribution would tend to lead to a closer approximation of "normal" adult skeletal proportions than would a more uniform reduction of growth rates' (Fleagle *et al.*, 1975). The second point is an important one: the arrangement is adaptive, in the biological sense of preserving the competitiveness of the species in the struggle for reproductive success. If we assume that the normal *shape* of cebus adapts it optimally to its ecological niche, then the response to chronic undernutrition, a circumstance clearly encountered very frequently, should be such that the advantages of normal morphology are preserved as closely as possible. What is sacrificed is size, what is preserved is shape. The same is seen in human malnutrition.

The results of rehabilitation have recently been reported. Catch-up growth in weight is rapid and complete in both groups, and by 22 weeks of rehabilitation (1 year of age) limb catch-up was complete in the calorie-deficient group and very nearly so in the protein-deficient group. Clearly size and shape in the end returned to normal.

These experiments had another and highly instructive side. In real life undernutrition of human infants almost invariably occurs in combination with social deprivation in conditions of poverty. Hence, when such infants are treated in hospital and returned to their families and re-tested months or years later, it is extremely difficult to know whether any deficit in intellectual or emotional functioning is due to the malnutrition itself or to the effect of the associated social circumstances. Even comparison with sibs who did not enter the hospital raises the obvious problem of whether one child is favoured over another.

The Harvard investigators tackled this problem by dividing the cebus infants on restricted diets each into two groups. One group was reared in partial isolation, in a solitary cage within sight and sound of other monkeys; and the other group was housed similarly but had 3 hours of play every weekday with 1 to 3 other monkeys in a large play-pen, plus daily human handling. The monkey analogues of intelligence and emotional poise in humans are perhaps not easy to define; exploratory behaviour was the item chiefly studied, plus motor behaviour, general activity and the occurrence of items such as rocking, clutching of the body and frowning, which signify emotional disturbance. By the end of the 20-week period of food restriction it was clear that well fed and socially deprived animals scored about the same as ill fed and socially enriched ones, both groups scoring lower in all items than the well fed and socially enriched. The authors, M. F. Elias and K. W. Samonds, concluded that their findings were consistent with many others in showing that the behavioural consequences of calorie malnutrition were in many respects similar to those of emotional deprivation. Simultaneous deprivation in both factors was more severe than deprivation in either alone, but the two factors were additive rather than interactive in effect. The retardation of behavioural development was less than that of growth in size, though probably little less than that of skeletal maturity. After six months rehabilitation, no behavioural deficit could be found in either of the two nutritional groups; recovery was complete.

From a practical point of view the outcome of the monkey experiments and of those human experiences which are carefully controlled seem to agree in suggesting that events subsequent to a period of malnutrition occurring either in late foetal or early postnatal life exert an overridingly important effect on its outcome. The restorative powers are very great provided social circumstances allow them full play.

Bigger not Better. One last important point needs to be made here. It should not be uncritically assumed that bigger means better. Just because in temperate industrialized countries the better-off members of the community are larger than the worse-off and tall women have more successful reproductive histories than short women, it does not follow that the same holds true under other ecological conditions. In the harsh environment of the Peruvian Andes, it is the small mothers who have more surviving offspring, as A. R. Frisancho and his colleagues have shown in the shanty towns around Cuzco. Small body size may be more adaptive under some conditions. L. Malcolm in New Guinea and W. A. Stini in Latin America have made the same point. In an agricultural peasant economy a small man is more efficient than a tall one, requiring to do less work to feed himself. Level of nutrition has to be seen as part of the whole ecology, of a whole philosophy even. Overnutrition is no less lethal than undernutrition, and a great deal more prevalent in many parts of the world.

3. **Differences between Races**

Populations differ in their average adult size and their tempo of growth as well as in their shape. These differences, as we have seen, are due to a complicated interaction of genetical and environmental factors. In this section we concentrate on the effect of differences in gene pools represented by the rather arbitrary classification of race. We rely on geographical and historical origins for our criteria for these major population groups, and refer to them simply as European, African and Asiatic (by which last is meant Chinese, Japanese and Indo-Malayan). For the purposes in hand we want to compare representative groups of each race growing up, so far as possible, under similar and preferably optimal environmental conditions. Extensive relevant data may be found in the International Biological Programme book *World-wide Variation in Human Growth* by Dr Phyllis Eveleth and the present writer (1976).

Figure 42 shows the height curves of London children, well-off Chinese children in Hong Kong, and Afro-Americans from Washington, D.C., who, though from a relatively low-income group in the United States, enjoyed a more favourable environment than any African group studied in large numbers in Africa (the Nigerian well-off group approach this condition, however; see p. 130). The European and Afro-American groups have almost identical curves for boys; the African-descended girls, however, are a little larger than the European girls. This is partly due to their having a swifter tempo, with menarche about 0·3 years earlier.

The Asiatic boys and girls are distinctly less tall, despite coming from high socio-economic groups receiving better-than-average care. It is not growth delay that makes them smaller; indeed their

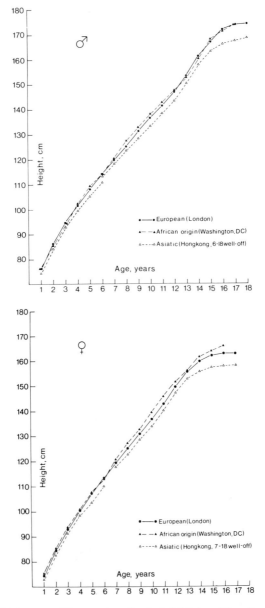

FIG. 42 (*a*) Height means of European boys (London), Asiatic boys (Hong Kong) and U.S.A. boys of African origin (Washington, D.C.); (*b*) Height means of European girls (London), Asiatic girls (Hong Kong) and U.S.A. girls of African origin (Washington D.C.) (From Eveleth and Tanner, 1976)

tempo of growth is significantly faster than that of Londoners, with menarche at 12·5 compared with 13·0 years. Hence the adult height difference is greater than the childhood difference, not less. This difference is due mainly to differences in gene pool. A frequently-quoted finding, due to Professor William Greulich of Stanford University, is that Japanese children growing up in Los Angeles are taller than Japanese growing up in Japan. This was indeed the case in the 1950s, but by now the marked trend towards increasing height in Japan has eliminated the difference. Japanese in Japan, Hawaii and California no longer differ significantly in height; but both differ from Hawaians and Californians of European or African descent, their mean heights being at about the 15th centile of the British standards.

Bodily Proportions. The largest differences between races, when all are growing up in good environments, are those of shape. Figure 43 shows lines representing sitting-height means at successive ages plotted against leg-length means at the same ages, in boys. The London boys (solid line) show a straight line up to adolescence, at which time sitting height spurts more than leg length, causing the line to bend upwards. The Chinese when small have a similar body proportion, but by adolescence their sitting height has become considerably greater for a given leg length.· The Africans, on the other hand, have a much lower sitting height for leg length (or,

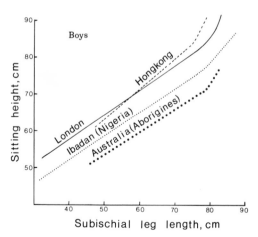

Fig. 43. Regression of sitting height means on leg length means at successive ages of European, Asiatic, African and Australian populations
(From Eveleth and Tanner, 1976)

equivalently, longer legs for similar sitting height) and the Australian Aboriginals have still longer legs than the Africans. These are characteristic differences, shown by females as well as males (see Hamill and colleagues, 1973, and Greulich's paper). Figure 44 illustrates the European–African differences in relation to two typical Olympic sprinters.

An equally characteristic difference between European and African is in the relation of shoulder width to hip width. In both sexes the African has slimmer hips for a given shoulder width. The proportions of the Asiatic in this respect are similar to those of Europeans. There are differences in body composition also, Africans having more muscle and heavier bones per unit weight, at least in males, together with less fat in the limbs in proportion to fat on the trunk (Johnston *et al.*, 1974; Eveleth and Tanner, 1976). These differences confer distinct advantages and disadvantages in

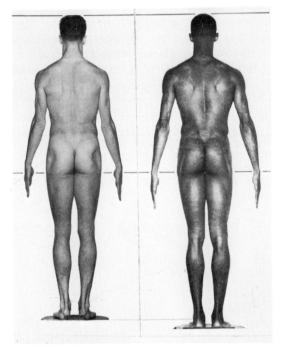

FIG. 44. Comparison of European and African physiques with photographs enlarged so that both have the same sitting height: two Olympic 400 m. runners
(From Tanner, 1964)

certain sports, with results that can be seen in Olympic records (Tanner, 1964). Africans have an advantage in many track events, especially the high hurdles; the Asiatics in gymnastics and weight-lifting.

Tempo of Growth. The African newborn is ahead of the European in skeletal maturity and motor development. He maintains this advance only for some 2 or 3 years in most areas in Africa, but this is because nutritional disadvantage supervenes. In America and Europe the African stays in advance, in bone age and also in dental maturity. In a nationwide survey of the United States, in which a true proportionate sampling of the whole population was attempted, the mean ages of menarche were 12·5 years for African-descended and 12·8 years for European-descended. Well-off Asiatic groups have as fast a tempo as Africans, in later childhood if not in earlier years. Mean age of menarche in Hong Kong girls from affluent families was found by Francis Chang and his associates to be 12·5 years. In Japan in the 1960s a large sample of girls from several urban areas gave a mean value of 12·9 years.

4. Climatic and Seasonal Effects on Growth

It might be argued that the differences in growth that occur between Europeans, Africans and Asiatics might be due to the direct effect of climate on the growing child, were it not that we are able to compare descendants of Africans and Europeans living in the same area. The differences in growth have indeed been brought about in response to the different ecological conditions in which each group evolved, but they arise from selection over many generations rather than from the immediate effect on individual children. The long limbs of the African enable him to lose more heat per unit volume than the European, and the thick-set body and short limbs of the northern Asiatic are similarly adaptive in arctic regions. There is, in fact, quite a close correlation between the linearity of peoples, as judged by their weight per unit height as adults, and the average annual temperature of the area where they live, or from which they migrated in historical times. Thus over thousands of years children and adults with genes leading to climate-adaptive characteristics have survived to breed more frequently than others.

Only one direct effect of climate on growth is certainly known. The high altitude of the Peruvian *altiplano* (4,000 metres) induces a larger chest circumference and bigger lungs in Quechua children growing there than occur in Quechua children on the sea coast.

One alleged climatic efféct has achieved a certain notoriety; nine-teenth-century medical textbooks stated that menarche was earlier in girls living in the tropics, and this information has been duly copied from one textbook to another till quite recently. In fact, modern statistics makes it clear that climate has at most a very minor effect on age at menarche; people living in tropical countries mostly have, in fact, a late menarche, but simply because their nutritional level is low. Well-off children in these areas have menarche at approximately the same time as children in temperate zones, and children of temperate-zone parentage growing up in the tropics have menarche at a normal time for the population from which they are drawn. A list of median ages of menarche in different populations is given in Table 3.

Season of the year, however, exerts a considerable influence on velocity of growth, at least in west European children. Children grow faster in height in spring and summer than in autumn and winter. The writer's colleague Dr W. A. Marshall found that in children aged 7 to 10 years maximum height velocities were reached in the 3-month periods ending between March and July. During the 3 months of fastest growth a child on average grew three times as much as during his 3 months of slowest growth. Thus the difference is considerable and sufficient to make it important to study growth over a whole year when comparing the effect of a hormone or a drug on growth rates.

Marshall also showed that totally blind children failed to synchronize their growth rate variations with the time of year, though the variations themselves occurred to the same degree as in sighted children. The cause of the variations is unknown; but whatever it is, light or some radiation falling on the eye 'entrains' them to season of the year (as it entrains the breeding-cycle of many animals).

In tropical countries seasonal variations tend more to be governed by the rainy and dry periods, and follow food supply and the frequency of infections.

5. Effects of Diseases

In well-nourished children the effects on growth of minor diseases such as measles, antibiotic-treated middle ear infection, or even pneumonia, are minimal.

In children with less adequate diets disease may cause a slowing down of growth, followed by a catch-up when the disease is cured. Often children with continual colds, ear disease, sore throats and

TABLE 3

Europe		Near East and India	
Oslo	13·2	Bagdad (well-off)	13·6
Stockholm	13·1	Istanbul (well-off)	12·3
Helsinki	13·2	Tel Aviv	13·2
Copenhagen	13·2	Iran (urban)	13·3
Netherlands	13·4	Tunis (well-off)	13·4
North-east England	13·4	Madras (urban)	12·8
London	13·0	Madras (rural)	14·2
Belgium	13·0		
Paris	13·2	*Asiatics*	
Zurich	13·1	Burma	13·2
Moscow	13·0	Singapore (average)	12·7
Warsaw	13·0	Hong Kong (well-off)	12·5
Budapest	12·8	Japan (urban)	12·9
Romania (urban)	13·3	Mexico	12·8
Carrara, Italy	12·6	Yucatan (well-off)	12·5
Naples (rural)	12·5	Eskimo	13·8
European-descended		*Africans*	
Montreal	13·1	Uganda (well-off)	13·4
USA, all areas	12·8	Nigeria, Ibadan	13·3
Sydney	13·0	(university-educated parents)	
New Zealand	13·0	South Africa (urban)	14·9
Pacific		*African-descended*	
New Zealand (Maori)	12·7	USA, all areas	12·5
New Guinea (Bundi)	18·0	Cuba, all areas	13·0
New Guinea (Megiar)	15·5	Martinique	14·0

Mean ages of menarche (years) in various population groups. All data refer to period between 1960 and 1975: status quo data with means calculated by probits or logits.
(Sources of data will be found in Eveleth and Tanner, 1976 and Oduntan *et al*. 1976)

skin infections are on average smaller than others, but they tend to come from economically depressed and socially disorganized homes, where proper meals are unknown and cleanliness too much trouble. The small size is more likely to be due to undernutrition than to the effects of continual minor diseases.

Major diseases may slow down growth, though the effects are seldom permanent. Actual disorders of the growth process are discussed briefly in Chapter 12.

6. Psychosocial Stress

There is now clear evidence that in some children psychological stress causes relative failure to grow (see Chapter 12). It does this by inhibiting the secretion of growth hormone. When the stress is removed, secretion of growth hormone resumes again, and in clinical cases a catch-up occurs which is indistinguishable from the catch-up following administration of growth hormone to a child otherwise permanently deficient in it.

Just how far information derived from such clinical cases may be extrapolated into the life of the ordinary child is hard to say. Fairly severe stress seems usually to be involved in these clinical cases and, of course, the majority of children given sufficient food continue to grow even in astonishingly stressful circumstances. In a famous experiment (originally planned to illuminate an entirely different issue) Dr Elsie Widdowson of Cambridge University furnished evidence that the presence of a sadistic schoolteacher caused a slowing down of growth in children in an orphanage, despite a concurrent increase in the amount of food eaten. Presumably activity increased, unless indeed stress put up the requirements of maintenance. Many years ago Drs G. E. Friend and E. R. Bransby found that in certain boarding-schools boys grew more slowly in term-time than in the holidays at home. The recent experience of R. H. Whitehouse and the present writer tends to confirm this. On the other hand, a boarding-school may provide the friendly atmosphere required for catch-up to a child whose growth has been stunted by an adverse home, as illustrated in Chapter 12.

7. Effects of Urbanization

Children in urban areas are usually larger and have a more rapid tempo of growth than children in the villages of the surrounding countryside. This apparently results from the provision of a regular supply of goods, of health and sanitation services, and educational, recreational and welfare facilities. Areas of high population density which lack such characteristics, as for instance the shanty slums of South America and Africa, do not show the same effect on growth as the orderly towns of the industrialized countries. In Finland, Greece, Austria, Romania and Poland, boys in the towns are between 2 and 5 cm. taller (depending on age and the country considered) than boys in the villages. Some statistics on age at menarche are shown in Figure 45.

In Poland a particularly interesting study was made by Drs

Age of menarche, years

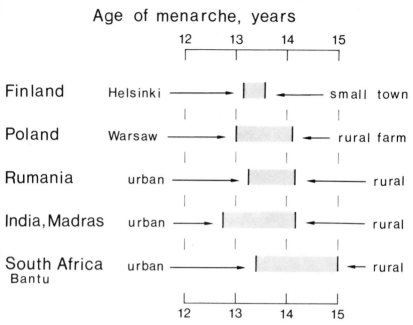

FIG. 45. Mean ages of menarche in urban and rural areas throughout the world. Shaded areas show difference between median ages in each geographic area.
(From Eveleth and Tanner, 1976)

S. Panek and E. Piasecki when a new town was founded, with the immigrants all coming from a near-by rural area. Children growing up in the town were compared with children whose parents stayed in the rural area. They averaged 4 cm. taller; in the girls menarche occurred earlier. It is possible that the migrants from country to town were a special selection of tall, fast-growing people. Such self-selection is characteristic of most migrant groups and probably usually occurs to a greater degree than in the Polish example. It is a well-known difficulty in all studies of voluntary migrants that they never seem to be a random sample of the population they leave, but are usually taller and more intelligent than the stay-at-homes (even if they migrate only from one English county to another).

Various highly speculative reasons have sometimes been given for the swifter growth of urban children. They range from more continuous illumination with artificial light to more continuous bombardment with sexual stimulation. In fact, the nutritional explanation suffices: rural children expend more energy in physical

Foetus into Man

activity than urban ones, and their total calorie intake is frequently less, despite townsmen's ideas to the contrary.

8. Effects of Socioeconomic Status and Numbers in the Family

Children from different socioeconomic levels differ in size and in tempo of growth in every society except one (p. 149) so far examined. The upper groups, defined usually in terms of father's occupation, are larger and have a more rapid tempo. In most European groups – both west and east – the chief difference is between those whose fathers are in occupations classified as 'manual' (or 'working') and 'non-manual' ('intellectual'). In developing countries similar or larger differences occur in relation to fathers' or mothers' educational status. The classification by occupation, as in the Registrar-General's classification in the U.K., has been very useful for many epidemiological purposes, but is becoming increasingly irrelevant as an indication of the standard of living and the child-centredness of the home, which are probably the operative factors.

The difference in height between children of professional and managerial fathers and those of unskilled manual labourers in the U.K. is currently about 2 cm. at age 3 years rising to 5 cm. at adolescence. In weight the difference is relatively slightly less, since the worse-off children have a higher weight for height because of a higher intake of sugar and other carbohydrates, relative to fat and protein. In the British National Child Development Survey – a nationwide sample of children consisting of all those born in the first week of March 1958 – an overall difference of 3·3 cm. was found between 7-year-old children of managerial and unskilled labouring fathers (Figure 46). Similarly in the U.S. national sample of children measured by the National Center for Health Statistics, children aged 6 to 11 were some 3 cm. taller in the rich families than in the poor ones. Some of this difference is due to a faster tempo of growth in the well-off. Some of the height difference persists into adult life, however, as E. Schreider and others have shown. At the University of Paris, the tallest students in the 1960s were those of parents in intellectual professions, while the shortest and heaviest were from worker and peasant families.

These differences are compounded by differences in height and tempo according to numbers of sibs in the family (Figure 46). As one might anticipate, first-born children are somewhat taller than later-born children with the same number of sibs, since they have

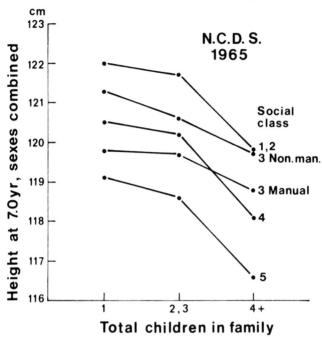

FIG. 46. Height at 7·0 years of British children (sexes combined) according to number of children in the family and socio-economic level (1–5)
(Data from National Child Development Study, 1965; given in Goldstein, 1971)

had a period of being an only child. Thus the effects of being a later-born child (for given age of mother) and of having younger sibs summate. The more mouths to feed, it seems, or simply children to look after, the more slowly the children grow. This difference is one solely of tempo, for brothers do not differ systematically according to birth order when they are fully grown. Professor Harvey Goldstein, previously at the writer's department and now at the London University Institute of Education, showed that in the N.C.D.S. data about half of the social class differences in height persisted even when the effect of numbers of sibs was allowed for. His analysis is the most thorough yet done of the complex and interacting factors affecting height at a specific age, and is highly recommended to readers with a knowledge of the statistics of the analysis of variance. D. F. Roberts and his colleagues have shown that in north-east England class differences in age at menarche, unlike those in height, do now disappear when the numbers of sibs are allowed for.

All the differences associated with father's occupation may not be of direct environmental origin. Social classes are to some extent endogamous, and movement from one to another in some cultures seems to be linked with body size as well as with intellectual ability and physical energy. In Belgium, R. L. Cliquet (1968) has shown that young men who were entering an occupation exceeding in prestige that of their fathers were taller as well as cleverer than those who stayed in the same or an equivalent occupation. A similar social movement occurred amongst women in Aberdeen in the 1950s. Girls who entered non-manual jobs were taller, as usual, than those entering manual ones. In addition, however, the separation was increased in selection of a husband. The taller girls, both in the manual and the non-manual occupations, more often married men in non-manual jobs, and the shorter girls more often men in manual ones (Figure 47).

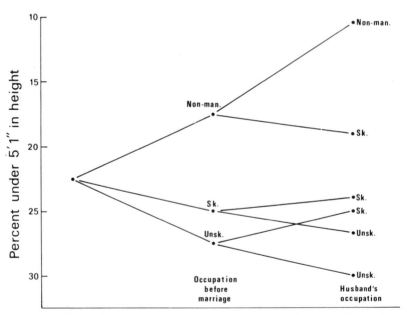

Aberdeen primiparae born to skilled manual workers

FIG. 47. Percentage of daughters of skilled manual workers under 5 ft. 1 in. tall taking non-manual and manual jobs and marrying men in non-manual, skilled manual and unskilled manual occupations
(Redrawn from Thomson, 1959)

The one exception to this occupational class-gradient in children's heights is constituted by the most recent data from Sweden, published in 1976 by Dr Gunilla Lindgren of the School of Education in Stockholm. In a survey of children in all urban areas of Sweden (though not, unfortunately, in rural ones) no difference according to father's occupation was found, either of height at a given age (between 7 and 17) or in age at menarche. In agreement with this, no difference in height according to father's occupation can now be demonstrated in Swedish men conscripted for military service. Perhaps children's growth provides a meaningful measure of the classlessness of society, if that famous phrase is defined in functional rather than rhetorical terms.

Height and Mental Ability. There is a curious relation between height and measurements of intellectual ability. In all countries students are uniformly the tallest group in the population, averaging 2 to 3 cm. taller than the mean of the total population (see Eveleth and Tanner, 1976). At the other end of the scale children in the U.K. National Child Development Survey designated as requiring placement in schools for the educationally subnormal were 3 cm. less tall at age 11 than others, even when matched for father's occupation. Those who were in ordinary schools but who teachers thought might benefit from such education were about 2·5 cm. less tall, and those receiving special help within normal schools were 1·5 cm. less tall, again computing the differences within occupational groups of the fathers (Peckham *et al.*, 1977). Part of this difference may be due to differences in tempo, but it is a common finding that mentally-handicapped adults are on average smaller than others. Why this should be so is not at all clear.

In schools there is a small correlation uniformly found between height and scores on tests of ability or intelligence, of the order of 0·3. Part of this is due to linked tempo differences in bodily and mental development. But part is not, for surprisingly the correlation seems to persist into adult life. In Swedish conscripts a correlation of 0·24 was found, and other data give similar values. Obviously such a small correlation is meaningless in the individual case, but it should tell us something about our culture. The situation is statistically comparable to accident rates, which do not tell an individual whether he personally will be killed at a particular cross-roads, but which nevertheless reflect accurately the state of the roads at that point.

9. Secular Trend

During approximately the last hundred years in industrialized countries, and recently in some developing ones, children have been getting larger and growing to maturity more rapidly. This is known as the 'secular trend' in growth. Its magnitude is such that in Europe, America and Japan it has dwarfed the differences between occupational groups.

Figure 48 shows Swedish data, which are the most comprehensive available for any country over the whole period. The differences in children's heights between 1883 and 1938 amount to nearly 1½ years of growth; differences between 1938 and 1968 are much less. At the age when growth ceases, as shown by the 16–18 year-old girls, the secular trend is still present, but much smaller than in childhood.

The occurrence of a secular increase in height and weight has been documented in nearly all European countries, including Sweden, Finland, Norway, France, the U.K., Italy, Germany, Czechoslovakia, Poland, Hungary, Soviet Union, Holland, Belgium, Switzerland and Austria. From about 1900 to the present, children in average economic circumstances have increased in height at ages 5 to 7 years by about 1–2 cm. per decade. The trend

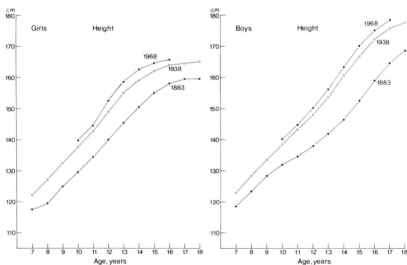

FIG. 48. Secular trend in growth of height – Swedish boys and girls measured in 1883, 1938–9 and 1965–71
(From Ljung, *et al.*, 1974)

starts early in childhood, as pre-school data make clear. At least in Britain it began a considerable time ago, for Charles Robert, a factory doctor writing in 1876, said: 'A factory child of the present day at the age of nine years weighs as much as one of 10 years did in 1833 . . . each age has gained one year in forty.' In 1833, the first relatively largescale measurement of children took place to provide parliamentary evidence on the effect of the employment of children in factories. At that time, working boys aged 10 years averaged 121 cm. in height compared with 140 cm. today; those aged 18 years averaged ·160 cm. compared with 175 cm. today. These differences are actually a good deal larger than the differences seen at present between urban slum-dwelling children in the underdeveloped countries and the affluent children of the industrialized West (see Figure 4).

The trend in Canada, the U.S.A., Australia and other countries has been similar. Japan shows a particularly dramatic trend; from 1950 to 1970 it amounted to about 3 cm./decade in 7-year-olds and 5 cm./decade in 12-year-olds. From 1900 to 1940 the trend was less than 1 cm./decade. In the industrialized countries the trend is now gradually stopping, as may be seen in the Swedish data. Dr Albert Damon, of Harvard University, showed that amongst American families with sons educated at Harvard in successive generations the increase in height was 2·6 cm. between the first two generations recorded (mean birthdates 1858 and 1888), 1·1 cm. between the second pair of generations (mean birthdates 1888 and 1918) and nil between the third pair of generations (mean birthdates 1918 and 1941). In poorer sections of the community, however, the trend continues.

During the same period there has been an upward trend in adult height but only to the lesser degree of about 1 cm. per decade since 1880. An astonishing series of Norwegian growth data stretching back to 1741 reported by V. Kiil indicate little if any increase in male height from 1790 to 1830, and these and other data show a trend of some 0·3 cm. per decade in several countries from 1830 to 1880, depending on their situation *vis-à-vis* the progress of the industrial revolution. As Dr van Wieringen of Utrecht University has shown in analysing the data on height of Dutch conscripts from 1851 to the present, periods of economic setback were associated with a stoppage or even a reversal of the secular trend.   At all times, however, the trend has been much smaller in adults than in children. Thus, much of the trend in children's heights is due to their

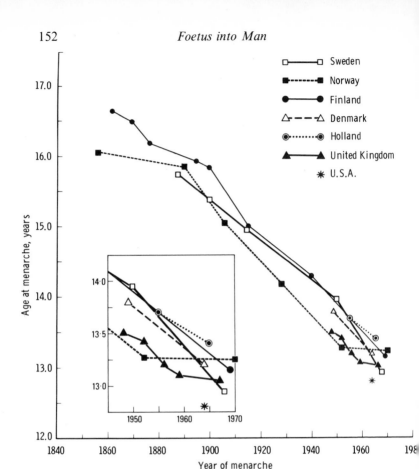

FIG. 49. Secular trend in age at menarche, 1860–1970.
(Sources of data and method of plotting detailed in Tanner, 1962, 1973)

maturing earlier. This is best shown by statistics on age at menarche, illustrated in Figure 49. The earlier data are all based on
recollected age, and are therefore suspect in detail. However, the
general trend is plain to see, and from 1880 to 1960 it averaged about
0·3 years per decade. Recently, in some places such as Oslo and
London, the trend has slowed down or stopped. In other areas, for
example parts of Holland and Hungary, it is still continuing. A list of
present-day ages of menarche in various countries, based on modern
probit-fitted data is given in Table 3 (see p. 143).

The causes of the trend are probably multiple. Certainly better
nutrition is a major factor, and perhaps in particular more protein

and calories in early infancy. A lessening of disease may also have contributed. Some authors have put forward the rather fanciful idea that increased psychosexual stimulation consequent on modern urban living has contributed, but there is no positive evidence for this. Girls in single-sex schools have been studiously compared with girls in co-educational schools in Finland and Sweden, with totally negative results, though whether this is a fair test of differences in psychosexual stimulation may be open to debate. Climatic changes have also been suggested. It is true that the world mean surface temperature rose from 1910 to 1940, but since then it fell again, to reach 1910 levels by 1970 (Lamb, 1973). However, climate seems to exert at most a very minor effect.

An additional explanation has been suggested for the increase in adult size, though not for the increase in tempo. It may be that some degree of dominance occurs in some of the genes governing stature; this would mean that when a person carrying mostly genes making for smallness marries a person carrying mostly genes making for tallness the height of their offspring would not lie on average exactly halfway between, but a little to the tall side. In more technical language, dominance is said to exist when some of the heterozygotes influencing stature produce effects nearer to the 'tall' than to the 'short' homozygotes. If this occurred (and only if) it would give rise to the phenomenon known as heterosis (or, in animals, hybrid vigour). An increased degree of outbreeding (i.e. of marriage to quite unrelated persons rather than cousins of various degrees) increases the number of heterozygotes, so if dominance did occur, then adult stature would be raised by the heterosis effect. There is increasing evidence that some degree of dominance in height genes, even if minor, may indeed exist. As for outbreeding, that has been steadily increasing ever since the introduction of the bicycle.

CHAPTER 10
The Organization of the Growth Process

We are now in a position to consider some basic principles that seem to characterize the growth process. We touched briefly on this topic in Chapter 2 when discussing cells as the basic unit of growth, but here we shall deal with the growth of the organism as a whole. Most of the examples are already familiar from earlier chapters.

Canalization and Catch-up
In Chapter 9 we saw that the birth lengths of monozygotic twins differed more than the birth lengths of dizygotic twins. However, by the time a few months had gone by, the reverse was the case. Those monozygotic twins who had suffered a poor position in the uterus had recovered from it and were back on their own trajectories of growth.

This is an example of what Waddington, in his classic book *The Strategy of the Genes*, called *canalization*. By this he meant that in any young animal growth has a tendency to return to its original path or channel if circumstances have conspired to push it off course. The processes of growth and differentiation are self-stabilizing or, to take another analogy, 'target-seeking'. The passage of a child along his growth curve can be thought of as analogous to the passage of a missile directed at a distant target. The target is determined by the genetic structure; and just as two missiles may follow slightly different paths but both end by hitting the target, so two children may have slightly different courses of growth but end up with almost the same physique. This self-correcting and goal-seeking capacity was once thought to be a very special property of living things, but now we understand more about the dynamics of complex systems consisting of many interacting substances we realize that it is not, after all, exceptional. Many complex systems, even of quite simple lifeless substances, show such internal regulation simply as a property consequent on their organization.

The power to stabilize and return to a predetermined growth curve after being pushed, so to speak, off trajectory persists throughout the whole period of growth and is seen in the response

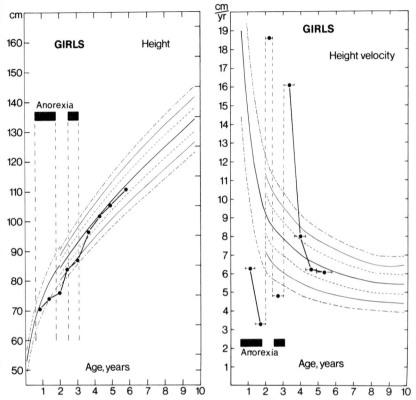

FIG. 50. Length and length velocity curves of infant with two periods of undernutrition (anorexia) to show catch-up
(Redrawn from Prader, Tanner and von Harnack, 1963)

of young animals to illness or starvation. During starvation an animal's growth slows down, but when feeding begins again its velocity increases to above normal for its age or maturity. Unless the starvation has been prolonged or has occurred very early in life, the original growth curve is caught up to and then once again followed. Figure 50 gives an example in a young baby who had two periods of starvation; other examples are shown in Figures 79 and 80 (pp. 212 and 216).

This rapid growth following the end of a period of growth restriction (for whatever reason) has been named *catch-up growth*. (It has also been called 'compensatory growth' by animal nutritionists, but this phrase has been pre-empted by zoologists to designate in mammals the excessive growth of, for example, the second kidney, when the first is damaged or removed, and in amphibians the re-growth of whole limbs following limb removal.) Catch-up growth may be complete and restore the situation, so far as we can tell, entirely to normal. There are two ways in which this may be done (Figure 51). In true complete catch-up the velocity increases to such an extent that the original curve is attained and thereafter growth proceeds normally (curve A in the figure). In complete catch-up with delay the catch-up velocity is not sufficient to do this, but maturity is delayed and growth is resumed at the correct velocity for chronological age (curve C) or, more usually, for maturity as signified by bone age (curve B). In these cases growth con-

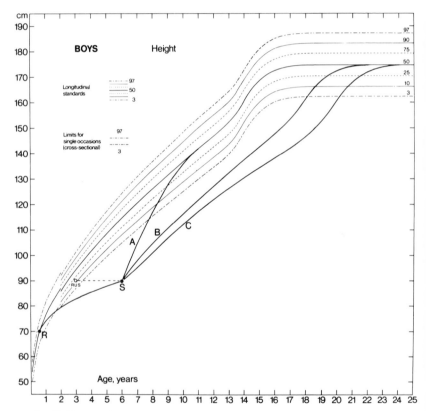

Fig. 51. Diagram of two sorts of complete catch-up. Growth retardation occurs from R to S with rehabilitation commencing at S. True complete catch-up A; complete catch-up by prolonged growth B and C. In C growth is resumed at average velocity for chronological age; in B at average velocity for bone age, marked R U S.

tinues for longer, but in the end complete catch-up, or compensation for the growth arrest, is achieved.

Frequently the catch-up is a mixture of these responses. In Figure 52 are shown the height curves of two brothers both of whom had isolated growth hormone deficiency from birth. The elder was not treated till age 6·2 years. By this time, besides being small, he was much delayed in growth, with a bone age of 3·0 'years'. He had a marked catch-up velocity, shown in Figure 53, but it was insufficient to bring him within the normal range of height-for-age curves.

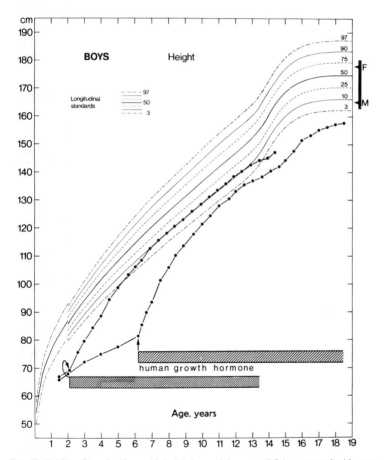

FIG. 52. Heights of two brothers with isolated growth-hormone deficiency treated with HGH from ages 6·2 years and 2·1 years. Catch-up of older brother is partly by high velocity and partly by prolonged growth and is incomplete. Younger brother shows true complete catch-up. F and M, parents' height centiles; vertical thick line, range of expected heights for family.

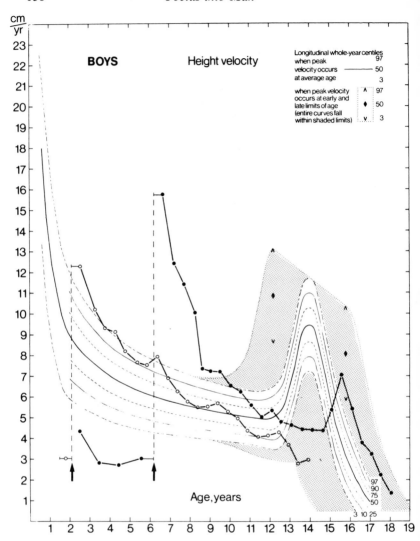

FIG. 53. Height velocities of brothers shown in Fig. 52. Arrows signify beginning of HGH treatment.

He remained somewhat delayed and his adolescent spurt took place some eighteen months later than average. Even so he ended only a little below the range of variation expected from his parents' heights (shown in the figure by the vertical thick line with M and F, representing mother's and father's centiles, upon it). His younger brother, whose treatment started much earlier, at age 2·1 years, showed a classical complete catch-up, reaching the 10th centile after 5 years of treatment and thereafter continuing along it (notice this is his mother's centile).

Children have an astonishing capacity to return to their growth curves, and a 5-year-old whose growth has been quite arrested for a whole year, e.g. by hypothyroidism (see Chapter 12, pp. 214–15) will catch up completely provided he is well fed and looked after during his period of rehabilitation. Pathology or undernutrition early in life is another matter and children starved early in utero through some imperfection of the placenta usually fail to catch up completely. We do not know for certain why this is so, but it is thought that it relates to the phase of cell multiplication. As explained in Chapter 2, it seems that cell division has to take place to a tightly organized schedule and cannot be delayed while the organism waits for better conditions, as can the later phase of packing material into the cells.

An alternative or more probably a complementary explanation relates to one of the few extant theories of the catch-up mechanism. This is based on the fact that the child's catch-up velocity is large to start with and decreases progressively as the target curve is approached. It looks, therefore, as though the velocity is regulated by a signal of mismatch between the ideal or target curve and the actual situation. If so, both must be in some way monitored. The theory proposes that the ideal, genetically programmed, curve is represented in the brain (probably the hypothalamus) by a process or tally (perhaps the accumulation of specific substances), which continues almost regardless of outside events. The actual situation may be signalled by molecules 'spun off' each time a molecule of protein is synthesized in cells of muscles, cartilage or even liver, or perhaps when cell division occurs in the generative zone of the cartilage cells. These molecules would interact with the brain tally, cancelling it. The growth velocity would be proportional to the uncancelled amount of the tally.

This is a highly speculative scheme, possibly quite wrong and certainly incomplete. At least, however, it provides a framework for thought and experiment. It explains the difficulty of compensating for early prenatal growth failure by supposing that the brain process representing the genetic programme can be affected by lack of the proper supply of molecules at the time when the process is being constructed. This would be in early foetal and mid-foetal life.

If it occurred, then the target itself would have been lowered and catch-up would henceforth be to this lowered target, instead of to the genetically programmed one. Later, when the neurons had finished forming or perhaps finished some aspect of their protein synthesis, the target would become more nearly inviolable.

The actual cause of the greater-than-usual velocity in catch-up is not known. It is not the result of greater-than-normal amounts of growth hormone. Just possibly it may be due to greater amounts of somatomedin, but more likely to greater receptivity of the cartilage and other cells, perhaps brought about by some generally circulating messenger, for all cells of the body join in the catch-up spurt. Our ignorance about the catch-up mechanism is not too surprising when we reflect that we do not even know why younger animals grow faster than older ones.

One thing does seem to be clear about canalization. Regulation is better in females than in males. This applies to man as well as to other mammals. The radiation of the atomic bombs in Japan, for example, slowed down the growth of exposed boys more than that of girls, and the same is generally true of undernutrition except in social situations where the male is clearly favoured. Similarly, girls recover from growth arrest more quickly than boys. The physiological reason for this greater stability is not known.

Competence and Specification

All cells have a full set of chromosomes and hence of potential assembly-lines for all types of protein (see Chapter 2, pp. 26–7). But very early in embryological development this general competence, as it is called, is lost; each cell concentrates more and more on certain specialized production-lines. It seems that the other lines are put into mothballs rather than actually smashed up, but the mothballing is usually pretty thick. Thus when the period of competence of a cell to do something has passed, it is usually impossible to reactivate it.

The nervous system provides a good example of this property. The developmental programme of nerve cells is under the control of genes. At each stage of development some genes are turned on, and others turned off, usually for ever. The regulator molecules which do the turning on and off may either be resident in the cells themselves or may diffuse out of the cells in which they are produced and regulate neighbouring cells. Thus they may exert differential actions on cells at different distances away or on cells in different positions in the embryo. In mammalian and amphibian development the diffusing molecule mechanism is the chief one; a field of differential distribution of regulating signals is set up and cells are labelled according to their position in this field.

The most thoroughly studied instance is that of the toad retina. There is a small area of the embryonic nervous system which will develop into the neural part of the retina. It is polarized very early into latent front and back ends, as can be shown by taking it out of the body and growing it in tissue culture. However, if it is excised while still an early eye-rudiment and replaced back-to-front it is able to change its polarity and a normal eye is formed. The change is made through interaction with the rest of the system surrounding it. But if the rudiment is excised, left to develop to a more advanced stage outside the body in tissue culture, and then reimplanted back-to-front it can no longer reverse polarity and the eye forms an inverted pattern of nerve connections with the brain. This time the neurons are said to be 'specified' or 'determined'. Complete specification seems to accompany what Jacobson calls 'withdrawal from the mitotic cycle', i.e. cessation of the power of dividing. The time at which specification occurs varies from tissue to tissue and even from cell to cell. Specification usually occurs long before any morphological signs are visible; thus an experiment may reveal that certain cells have been specified irreversibly to make a certain part of the brain; but only considerably later can they be identified as actually turning into that part. Specification, in other words, is a matter of molecular form and distribution. The effect of a central-nervous-system poison in the foetus (e.g. thalidomide) depends firstly on the specific chemical affinities of the poison with particular types of cell and secondly on whether these cells and their potential replacements are specified or not at the time when the poison acts.

Sensitive Periods

A very important extension of this principle is seen in the concept of *sensitive periods*. They were originally called 'critical periods' both by biologists and psychologists, and have also, in a more limited context, been referred to as 'periods of special vulnerability'. We have already met several examples: the period when, just after it begins to use its eyes, the kitten must receive the impact of the visual environment (p. 115); the immediately neonatal period when the male rat's brain must receive testosterone from the animal's testes (pp. 57–8); and the period in the 14th postmenstrual week when the human male foetus must receive male sex hormone for differentiation of the external reproductive organs (p. 55).

All these are critical periods in the ordinary sense of the word, but in the physiological sense they are periods of increasing sensitivity of a receptor to a highly specific stimulus, followed by decreasing sensitivity and eventually by total lack of response. In the case of the kitten, for example, closure of eyes for 3 or 4 days at the height of the sensitive period in the 4th week after birth gives the same effect as closure for a month at 8 weeks. In the case of the rat's brain-differentiation the sensitivity can be accurately measured by seeing what dose of administered testosterone suffices to bring about the effect in the castrated rat. No amount will do the job before birth; a large amount is necessary during the first 2 days after birth, and a much smaller amount during the subsequent 3 days. Then after about 2 further days the period is past and once again no amount of testosterone can make good the damage.

There is a familiar example in human pathology, that of rubella, or German measles, in the pregnant mother. Some children whose mothers have rubella between the 1st and 12th weeks of pregnancy are born with cataracts in the eyes, and other defects. Rubella after the 12th week has no effect, so the period is a critical one of vulnerability.

The physiological basis of sensitive periods is not known in detail, but the general outline is clear enough. In the case of the rat there must be either a rising and waning amount of testosterone receptor in the particular part of the brain concerned, or else a rising and waning affinity of receptor to hormone. What is not clear is what causes the timing of this transient wave of sensitivity. If we knew this we should know much about the process of growth, for it seems as if sensitive periods are universal in all stages, at least of early development. One wave forms, rises, falls; its fall sets off another and *its* fall, a third.

Sensitive periods seem to be less frequent during postnatal growth, although this may merely reflect our inability to detect them. If they do continue to occur, particularly in the brain, they would be of the greatest concern for educationalists. It is widely believed, on educational grounds, that periods of sensitivity to various environmental sensory inputs indeed exist. Certainly there is a formal analogy between 'readiness to read', for example, and competence of brain systems. It is possible that the connection will be shown to be more than formal when research on brain growth has sufficiently advanced.

Heterochrony and Disharmonic Development

The multitude of chemical reactions going on during differentiation and growth demands the greatest precision in linkage. Thus for normal acuity of vision to occur the growth of the lens of the eye has to be harmonized closely with the growth in depth of the eyeball, so that the point of focus of the light-rays lies exactly on the retina. It is small wonder that the success of this co-ordination varies and most people are just a little far-sighted or near-sighted. Again, it seems that many features of the face and skull are individually governed by genes which do not much influence the growth of other, near-by, features. But in general the parts of the face fuse to form an acceptable whole. In this case the final growth stages are plastic and variable, and in fitting together, for example, upper and lower jaws, forces of mutual regulation come into play which do not reflect the original growth curves of the discrete parts.

The regulative forces harmonizing the velocity of growth of one part with that of another do not always succeed. If the original genetic forces begin by being too unbalanced, as in Down's syndrome (see p. 209), normal development cannot occur.

Variation in the speed of development of different structures and functions (heterochronisms) underlie many individual differences in bodily structure. Examples are the longer arms and legs relative to trunk in men as opposed to women, or in Africans as opposed to Europeans.

This general point of view, derived from experimental embryology, clearly has possibilities for behavioural research, even though it would be unwise to press the analogy too far. Not all human behaviour is open to such delightfully simple explanations as that of *Anser oedipoidipus* studied by Konrad Lorenz. Domesticated strains of geese mature earlier than wild geese, and in the offspring of a wilder gander and a domesticated goose disharmonization occurs between sexual maturation and the earlier mother-following response. The mother-following response still remains operative when the sexual response appears, and the young male bird consequently insists on copulating with its mother. Since the wild father's sexual activity arises only later in the spring, it is unnecessary that he be killed first; he remains wholly indifferent to the drama.

The behaviour of children with precocious puberty provides an instructive example of an extreme, pathological degree of heterochrony, in which a fully developed endocrine system acts upon a less developed brain. Psychosexual advancement by no means keeps pace with endocrine development. The hormones need a mature brain equipped with adolescent experience to work on if adult sex behaviour is to occur. This is not to say that the hormones are without effect. In one of the best described cases to date, by Drs Money and Hampson of the Johns Hopkins Hospital Psychoendocrine Clinic, the psychosexual development of a boy with simple precocious puberty was at a level characteristic of his chronological age, but more energized than is normal. The boy was $6\frac{1}{2}$ years old, with a bone age of 15. He had begun to have seminal emissions at 5 and to masturbate at 6. He experienced numerous dreams involving kissing women all over the body, which he would relate only to a male interviewer, with the air of a 'roué narrating his escapades in all-male company'. He had no knowledge of copulation, and overt sexual behaviour towards women had never been a problem. Several women on the hospital staff, however, said they felt uncomfortable under his gaze, which carried a considerable message of seduction.

Growth Maturity Gradients

One way in which the organization of growth shows itself is through the presence of maturity gradients. One such is illustrated in Figure 54. In Figure 54b the percentage of the adult value at each age is plotted for foot length, calf length, and thigh length in boys. At all ages the foot is nearer its adult status than is the calf, and the calf is nearer than the thigh. A maturity gradient is said to exist in the leg, running from advanced maturity at the far end of the limb to retarded maturity at the end nearest the trunk. The word 'gradient' has arisen because of the supposed mechanism by which this occurrence takes place. It is thought that in the embryonic limb bud, before any differences between the three segments can be discerned, there must be differences in the concentration of some chemical substance. Thus a concentration gradient of the chemical substance leads eventually to a maturity gradient in developing physical structure.

Figure 54a illustrates the same gradient in the arm, together with the fact that girls are more advanced towards maturity than are boys, without this affecting the gradient within the limb.

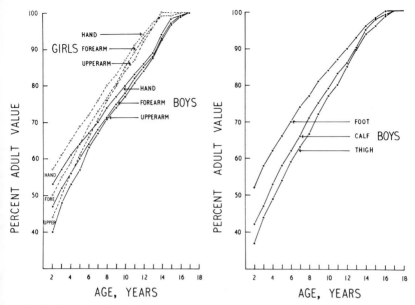

FIG. 54. Maturity gradients in length of: (*a*) upper limb; (*b*) lower limb
(From Tanner, 1962)

Many other growth maturity gradients are known, some covering small areas only and operating for short periods, others covering whole systems and operating throughout the whole of growth. The head is at all ages in advance of the trunk, and the trunk in advance of the limbs. In the hand and foot the second digit is the most advanced, and the third, fourth and fifth follow in that order.

Much of the growth of the brain is organized by means of such gradients and several have been described in Chapter 8.

Stages of Development: General and Singular
In child development there has been much argument as to whether development is continuous or whether it occurs in stages. Again, supposing stages do exist, there is argument as to whether any *general* ones occur, characterizing a simultaneous achievement of a number of anatomical, physiological and psychological developments.

In physical growth, there is continuity and little evidence for discrete stages. Even in the development of motor skills such as crawling and walking, development is more continuous than sudden. It is sometimes said that a child 'walks or does not walk', in a tone which implies that one day the child, previously unable to walk at all, suddenly finds the ability to walk unaided. Close observation of the development of such an ability, however, by no means supports this notion; on the contrary, the ability to walk develops gradually, as Dr Myrtle McGraw carefully documented many years ago.

In some creatures general stages of development do occur; e.g. in insects at metamorphosis. But in man only at adolescence do we have anything approaching a general stage; here, it is true, developments in anatomy, physiology and behaviour tend to occur simultaneously. Even here, however, the degree of synchrony is only relative. Skeletal age and dental age are largely independent of each other; and the spurt in bodily development is not matched by any great rise in intellectual capacity.

Physical development is best envisaged as a series of many successive processes, overlapping one another in time and linked loosely or tightly as the case may be. Out of the complexity of the linkages, under equilibratory forces, emerges an overall order in which visible changes follow one another with the regularity of a serially changing mosaic. The process is one of continuous unfolding and movement, with speeds varying from time to time in different parts of the mosaic; it is not a series of kaleidoscopic bumps. Only in certain restricted areas do rapid reassortments of the pieces occur, as they fall into new and increasingly precise patterns.

CHAPTER 11
Standards of Normal Growth

This chapter brings together, in a usable form, standards for normal growth and growth velocity in height and weight. It is a do-it-yourself kit for parents who want to know how tall their children will grow to be and for teachers or health visitors who want to see whether a child in their care is growing properly or not. As in all cases of do-it-yourself, the precision of the job depends on the care and understanding with which it is done. There is no inherent difficulty, except where x-rays are concerned.

The main standards presented here are derived from London children measured in the 1960s. For school-age children (5–16 years) a random sample of London schools was selected, and all pupils in them were measured, ensuring approximately 1,000 boys and 1,000 girls for each year of age. The pre-school sample was derived from all births registered in a central area of London but was much smaller, as tends to be the case in all longitudinal surveys. However, a current (1975–80) large-scale nationwide survey of English pre-school and school children (see Rona and Altman, 1977) shows these standards represent the present situation in England very well. They should be adequate for judging the growth progress of any child of European or African origin in a developed and relatively temperate country. With suitable caution they should apply also to children growing up under good social, educational and nutritional conditions in the developing countries. Children of Asiatic origin will be slightly smaller and have their peak velocities at a slightly earlier age (see Chapter 9). All the same, the reader who wishes to use these standards should be careful of two things. First and foremost, the lengths and heights *must* be taken with proper care ('garbage in, garbage out' is the auxologist's motto no less than the computer scientist's). Second, the various factors affecting growth described in Chapter 9 should be borne in mind in interpreting the results. For the benefit of North American readers the recent U.S. standards are also presented, on pages 188–91. However, these are for distance only, i.e. weight and height for age, and are based on cross-sectional data, which makes them more appropriate for assessing groups of children (as in public health surveys) than individual children.

It is worth remarking, perhaps, that collections of growth data may be used in three distinct ways. First, growth standards may be constructed to serve as a screening device – a very powerful one – for investigating groups of not overtly ill children to see which individuals might benefit from special medical, educational or social care (the *screening* or *community care* use). Secondly, growth standards may be constructed for studying the response to treat-ment of a child known to be ill (the *paediatric* use). Thirdly, growth data can be collected, and used as an index of the general health and nutrition of a population or sub-population. Such data can then be compared with simultaneously collected data from other populations or sub-populations (surveys) or with data collected on the same population or sub-population after the passage of some years (surveillance) or after some remedial, social or economic action. The population may be that of a whole country, or of a particular ethnic, social or occupational group within a country. For this use standards as such are not necessary; comparisons proceed by the classical statistical methods. In this chapter we are concerned only with the first two uses of growth data.

The best way to begin is by measuring and plotting (as if at age 19) the heights of father and mother and any grown-up brothers and sisters. The mother's height cannot be plotted unadjusted on the boy's chart or she will look ridiculously small. First the average difference between men and women should be added, which is approximately 13 cm. (5 ins.). In this way the mother's height *cen-tile* rather than her absolute height will be plotted. Similarly 13 cm. must be subtracted from the father's height before plotting it on a girl's chart. These plots done, the mid-point of the father's and mother's *centiles* is found and a line is drawn upwards from it for 8·5 cm. and another downwards for 8·5 cm. The band thus formed gives the range of height centiles within which 95% of the sons or daughters of the couple will be expected to fall as adults (Figure 55). (Strictly speaking it is the centile of the mid-parent height [average of father's and mother's heights] which should be plotted but the procedure described gives an acceptable approximation.) if the average parental centile is the 25th, the expected or 'target' range runs from below the population 3rd, to about the population 75th, centile. Such parents should not therefore expect all their children to be at the 50th centile, nor any to be at the 97th. Stan-dards which specifically allow for parents' heights appear later in this chapter (p. 195).

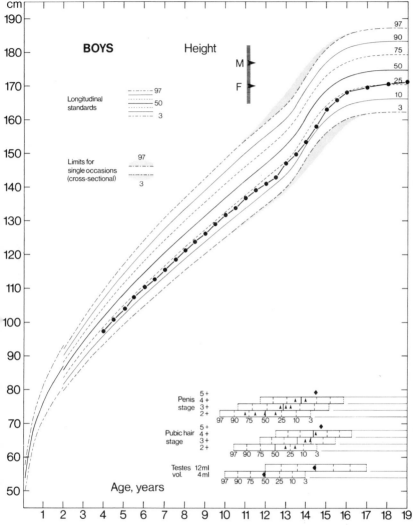

FIG. 55. Standards of height for boys, with normal boy plotted (solid circles). Open circles plot RUS bone age. M and F, parents' height centiles; vertical bar range of expected heights of offspring of these parents. Puberty ratings as shown

Figures 55 and 56 show the standards for height for boys and girls. These are longitudinal-type standards as explained on pp. 202–5, in which an average child having his growth spurt at the average time does actually follow the 50th centile. In Figure 55 an example is given, with parents' centiles plotted, and in Figure 56 the whole course of

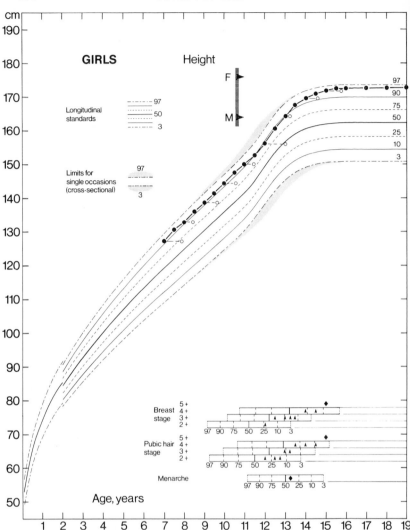

FIG. 56. Standards of height for girls, with normal girl plotted (solid circles). M and F, parents' height centiles; vertical bar range of expected heights of offspring of these parents. Puberty ratings as shown

growth of a normal girl from one of our growth studies has been plotted to show how such a complete record looks. The open circle against the first point plotted in Figure 55 represents the height plotted against the *bone age* instead of chronological age, so the length of the line from closed to open circles represents the bone-age delay in this boy. Bone age is a specialized investigation needing medical consultation (see Chapter 6, pp. 79–81). We shall consider its use later. The boxes in the lower right-hand portion of the charts relate to puberty standards and are also discussed later.

Decimal Age. The age scale has been calibrated in tenths of a year, not in months. This is designed to make things simpler, once the initial shock has worn off. When we calculate growth *velocity* we find it is straightforward if we use years and decimals of years, but horribly complicated (if done accurately) using years and months. Accordingly Table 4 gives the decimal calendar. The year is divided into thousandths and each date has a decimal number; thus 1 May 1975 is 75·329. The child's birthdate is similarly recorded; a child born on 23 June 1969 has the birthdate 69·474. Age at measurement is then obtained by simple subtraction, e.g. 75·329 − 69·474 = 5·855, shortened to 5·86 years. This is the value read on the age axis in plotting height for age. Suppose the next examination is done on 15 March 1976. This is the date 76·200; the child is 76·200 − 69·474 = 6·73, and the increment period is 6·73 − 5·86 = 0·87 years. Suppose the child has grown 4·9 cm. in this time. His growth rate would then be 4·9 ÷ 0·87 = 5·63 cm./yr and the plot would be made in the velocity chart (see p.) at the centre of the age interval, i.e. (6·73 + 5·86) ÷ 2 = 6·30 years.

Measuring Techniques

The measuring technique is all-important, especially if velocities are going to be considered, since an error in the direction of low values on the second occasion can actually have the child growing downwards. If the proper instruments and technique, shown in Figures 57 and 58, are unavailable, standing height can perfectly well be measured against the traditional door, with a triangular or book-shaped object brought down on to the child's head and the door marked. It takes at least two people to do this, however. One has to position the child, and another, at the right moment, brings the book, or, better, an architect's drawing triangle, down on the child's head. The longer side of the book (the spine) or of the triangle should be in contact with the wall, the shorter with the top of the head. The side in contact with the head must, however, extend well out, at least to a point above the forehead. The child stands with his heels against the bottom of the door; his shoulders, buttocks and head may touch the door, but that is not necessary. He looks straight forward (not nose-in-air) so that the earholes are

TABLE 4

Table of Decimals of Year

	1 JAN.	2 FEB.	3 MAR.	4 APR.	5 MAY	6 JUNE	7 JULY	8 AUG.	9 SEPT.	10 OCT.	11 NOV.	12 DEC.
1	000	085	162	247	329	414	496	581	666	748	833	915
2	003	088	164	249	332	416	499	584	668	751	836	918
3	005	090	167	252	334	419	501	586	671	753	838	921
4	008	093	170	255	337	422	504	589	674	756	841	923
5	011	096	173	258	340	425	507	592	677	759	844	926
6	014	099	175	260	342	427	510	595	679	762	847	929
7	016	101	178	263	345	430	512	597	682	764	849	932
8	019	104	181	266	348	433	515	600	685	767	852	934
9	022	107	184	268	351	436	518	603	688	770	855	937
10	025	110	186	271	353	438	521	605	690	773	858	940
11	027	112	189	274	356	441	523	608	693	775	860	942
12	030	115	192	277	359	444	526	611	696	778	863	945
13	033	118	195	279	362	447	529	614	699	781	866	948
14	036	121	197	282	364	449	532	616	701	784	868	951
15	038	123	200	285	367	452	534	619	704	786	871	953

	JAN. 1	FEB. 2	MAR. 3	APR. 4	MAY 5	JUNE 6	JULY 7	AUG. 8	SEPT. 9	OCT. 10	NOV. 11	DEC. 12
16	041	126	203	288	370	455	537	622	707	789	874	956
17	044	129	205	290	373	458	540	625	710	792	877	959
18	047	132	208	293	375	460	542	627	712	795	879	962
19	049	134	211	296	378	463	545	630	715	797	882	964
20	052	137	214	299	381	466	548	633	718	800	885	967
21	055	140	216	301	384	468	551	636	721	803	888	970
22	058	142	219	304	386	471	553	638	723	805	890	973
23	060	145	222	307	389	474	556	641	726	808	893	975
24	063	148	225	310	392	477	559	644	729	811	896	978
25	066	151	227	312	395	479	562	647	731	814	899	981
26	068	153	230	315	397	482	564	649	734	816	901	984
27	071	156	233	318	400	485	567	652	737	819	904	986
28	074	159	236	321	403	488	570	655	740	822	907	989
29	077		238	323	405	490	573	658	742	825	910	992
30	079		241	326	408	493	575	660	745	827	912	995
31	082		244		411		578	663		830		997

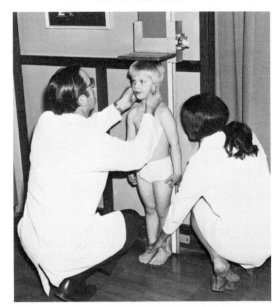

Fig. 57. Measurement of height   (Courtesy of R. H. Whitehouse)

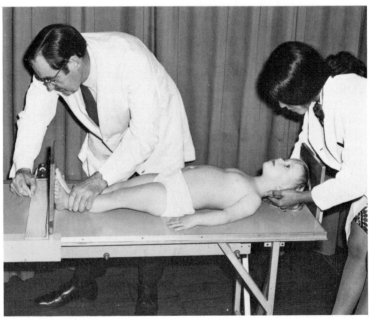

Fig. 58. Measurement of supine length   (Courtesy of R. H. Whitehouse)

in the same horizontal plane as the lower border of the eye sockets. He is told to reach up as much as he can, make himself tall, take a deep breath and relax his shoulders, and while this is going on the observer applies gentle pressure upwards to the bony prominences just behind the ears. The heels must not be raised from the ground and another observer may be necessary to ensure this. When the exact position has been reached the triangle is brought down on top of the head and the position of its corner marked on the door.

It will be appreciated how useful a counterbalanced headboard resting at all times on the child's head is, and how useful the direct digital read-out of the modern machines, the design of which we owe to the ingenuity of R. H. Whitehouse. The machine enables a single observer to take the measurement, provided only that he can trust the child not to lift his heels. Wavy metal rods attached to the backs of weighing-machines, or pieces of wood moving in ill-fitting metal collars, are hopelessly inaccurate and should be rejected in favour of the door-and-triangle approach.

It is much harder to find a home-made substitute for the supine length machine. The best that can be advised is a right-angled board, such as a bed with a low headboard and a plank put along it. The infant's head must be held as shown in Figure 58 with the ear–eye plane this time vertical. The ankles are gently pulled to stretch the child, the legs are straightened, the feet turned up vertically, the triangle brought against them and its position on the bed marked. Supine length averages about 1 cm. longer than standing height, but this difference varies from one child to another, its range running from 0 to 3 cm. In clinical work supine length is taken till age 2·0 years, and standing height thereafter, and the charts are constructed in this sense. From 2·0 to 3·0 years, however, both measurements are done, partly so that the difference can be evaluated and partly to ensure true velocities over this period, with each measurement done the same way on the two successive occasions.

Interpretation of Plot

The centiles shown in the standards have been explained in Chapter 1, p. 19. Height is Normally distributed, so the 50th centile corresponds to the mean of the population. The 3rd and 97th centiles, below and above which 3 in every 100 normal children lie, are the conventional limits of normality, or, more truthfully, of suspicion. We have to remind ourselves that 3% is not a negligible number. If every child under the 3rd centile in London came to the Growth Disorder Clinic, Great Ormond Street would be as crowded as Lourdes. Really pathological children lie a good way below the 3rd centile line.

Longitudinal versus Cross-sectional Standards

The standards in Figures 55 and 56 represent the curves of children who have their adolescent spurts at the average time. If a child is of average length at birth, has an average velocity throughout childhood, has the peak of his adolescent height spurt at the average time and ceases growth at the average age for doing this, then he will follow the 50th centile line exactly. If he has an early adolescent spurt his curve will leave the 50th to rise a little above it and then join it again later; if his spurt is late he will fall briefly below. In neither case will the deviation be very great. However, these are longitudinal-type standards, suitable for following the growth of individual children. The older cross-sectional type standards, which are not valid for individuals at puberty for the reasons discussed in Chapter 1, p. 10, show a wider variation. The 3rd and 97th centile of such standards are represented in Figures 55 and 56 by the outside of the shaded areas. It is logical to regard them as the limits of normality when a child is seen for the first time; but if the child's plot does fall in the shaded area, then by the second occasion of measurement, if he is normal, it will have moved towards the corresponding centile of the longitudinal standards, i.e., to within the non-shaded centile area.

Velocity Standards

More important than the child's exact position in relation to the 3rd centile in the distance chart is his plot in the chart of velocity. Most children with growth problems have a low velocity of growth. Velocity picks out the pathological cases far better than does distance (i.e. height for age). Velocity represents what is happening *now*, whereas distance represents the sum of all that has happened in the past. Thus a child who at age 6·0 years was at the 50th centile for distance and failed entirely to grow for a year would still be well within normal limits for distance at age 7·0. A study of his velocity from 6 to 7 would at once show him to be totally abnormal.

Figures 59 and 60 show the velocity standards for boys and girls. These are whole-year velocity standards, that is they represent velocities not only computed in centimetres per year, but measured

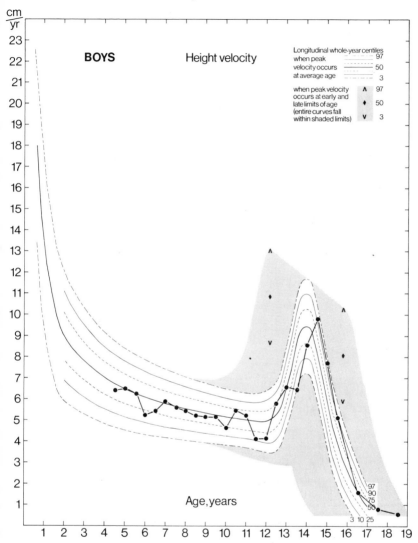

FIG. 59. Standards for height velocity in boys, with normal boy of Fig. 55 plotted over whole-year period, with new period commenced every 6 months (From Tanner and Whitehouse, 1976)

over approximately a year also. If a shorter interval is used then the variation amongst normals is greater. This is partly because the two errors of measurement assume a relatively greater importance in comparison with the actual growth that has occurred, and partly

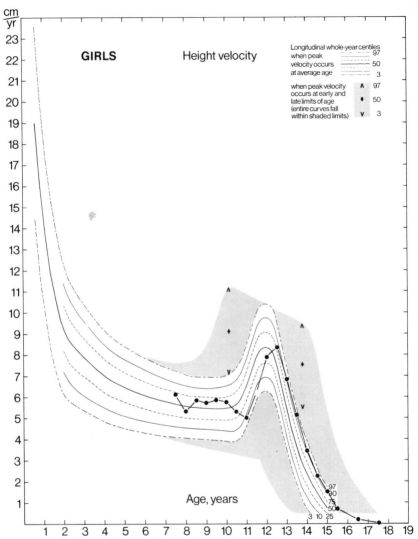

FIG. 60. Standards for height velocity in girls, with normal girls plotted over whole-year periods, with new period commenced every 6 months
(From Tanner and Whitehouse, 1976)

because of the seasonal influence making growth usually faster in the spring. For judging shorter periods than a year, therefore, the centiles should be farther apart. The girl represented in Figure 58 is plotted over whole-year periods, but a new yearly period is begun each 6 months, representing the sort of rolling average much used by economists (e.g. when the yearly rate of inflation is computed at the end of each month).

The velocity centiles at adolescence represent the situation for children all of whom have their peak height velocity at the average age for peak height velocity; i.e. 12·0 years in girls and 14·0 years in boys. Such children will follow one of the centiles during the spurt. In these charts the shaded portions represent the situation for early and late maturers. Boys who have their peak height velocity 1·8 years early (1·8 years being 2 standard deviations or the equivalent of the 3rd or 97th centile) have higher peaks and are represented by the diamond (50th centile) and the two arrows (3rd and 97th centile) at 12·2 years. Boys with a similarly late spurt are represented by the marks at age 15·8. All normal boys' velocity curves should fall within the shaded area, and have a shape approximating that of the centile lines between the areas.

One point of warning about velocity curves must be given. In distance curves a child tends to stay in the same centile throughout his whole growth, on average. But in velocity he cannot do so, unless his velocity is actually at the 50th centile. A child whose velocity is always at the 10th centile ends up pathologically small; and one whose velocity is consistently at the 80th ends up pathologically tall. Thus the pattern of an individual's velocity points throughout childhood is not the same as the pattern of his distance points.

Weight Charts

The weight distance charts are given in Figures 61 and 62 and the weight velocities in Figures 63 and 64. Delaying their appearance till now is deliberate. Because it is easy to measure, weight has come to have a quite disproportionate importance in the minds of many people. But because weight represents the sum of so many different tissues, it is harder to interpret a weight chart than a height one, and in the growth-disorder clinic weight charts are much less important. Weight may be a good measure of obesity, if, and only if, it is considered in relation to height; but skinfold thickness (see Chapter 1, p. 17) is a much better one.

Weight changes are important as an index in acute illness or starvation, of course, but in chronic long-term growth disorders height is more to the point. Weight can be very misleading sometimes. For example, a child with growth hormone deficiency

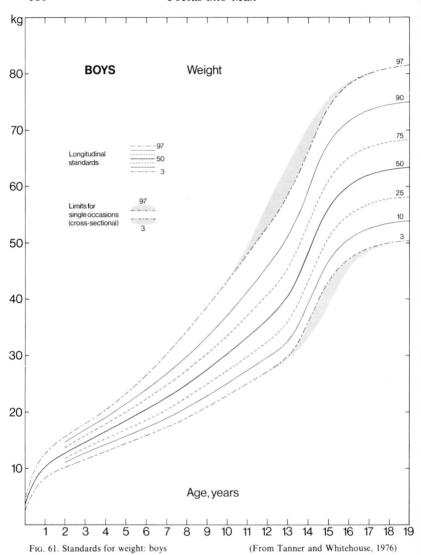

FIG. 61. Standards for weight: boys   (From Tanner and Whitehouse, 1976)

becomes fat as well as small, and when treated with human growth hormone he loses fat and gains height. In consequence, he may actually lose weight to begin with. Similarly when treatment is stopped he gains weight, but this is no sign of continuing growth; on the contrary, growth in height ceases.

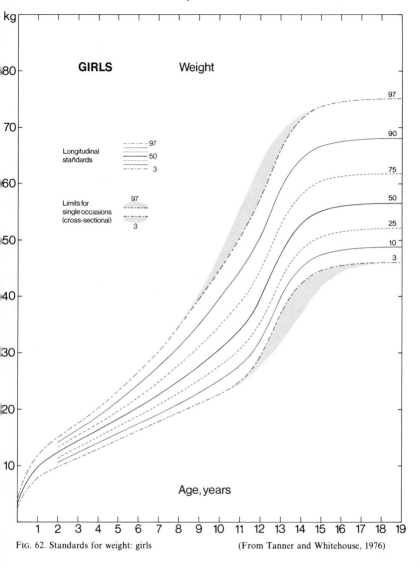

FIG. 62. Standards for weight: girls   (From Tanner and Whitehouse, 1976)

Charts of weight for given height, covering all ages mixed together, are often used by nutritionists in surveys (see Figure 71). They are attended by certain problems of too technical a nature to be gone into here. A good idea of a child's relative weight for height may be obtained, however, by comparing his centile positions in the

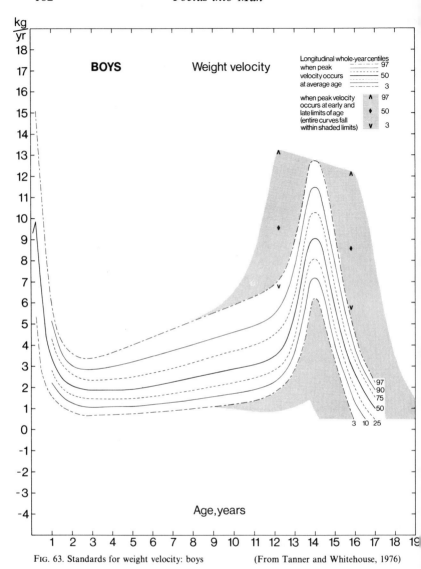

FIG. 63. Standards for weight velocity: boys   (From Tanner and Whitehouse, 1976)

two charts. At most ages a person at the 50th centile for height should be within the limits 10th and 90th for weight. A person at the 75th centile for height should be within somewhat higher limits for weight, very roughly given by moving all the centile lines in the weight chart upwards so that the 50th lies at the printed 75th, and

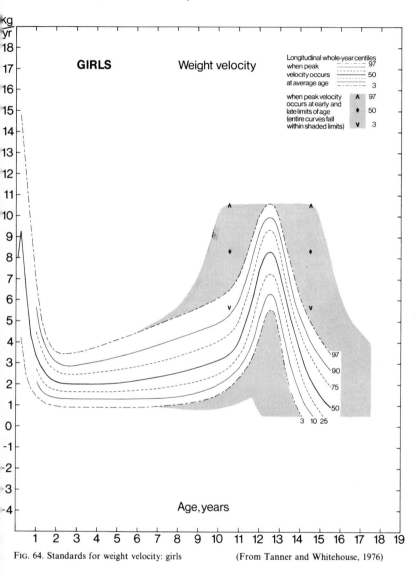

FIG. 64. Standards for weight velocity: girls   (From Tanner and Whitehouse, 1976)

then taking the 10th and 90th in this position. (For a person at the 90th centile for height the weight limits are shifted farther, but the limits become less accurate. It is then necessary to switch to a system, itself approximate, of standard deviation scores, which carries us beyond the scope of this book.)

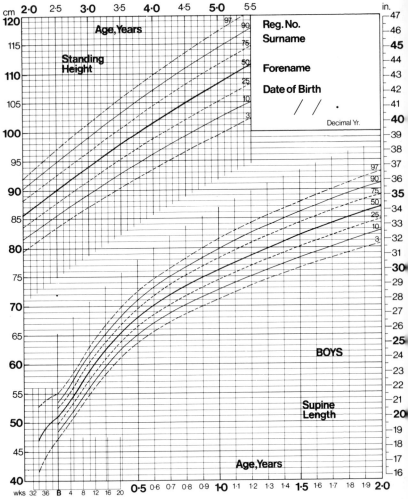

FIG. 65 (a). Standards for length–height; ages 0–5: boys
(From Tanner and Whitehouse, 1973)

Charts from Birth to Five Years

The charts for height and weight so far presented are not altogether convenient for young children because there is little room for detailed plotting. Figures 65 and 66 are consequently recommended for use as distance charts in this age group. They also have the advantage that children born pre-term can be accommodated on them. A child born at 36 weeks instead of 40 should not, 6 months

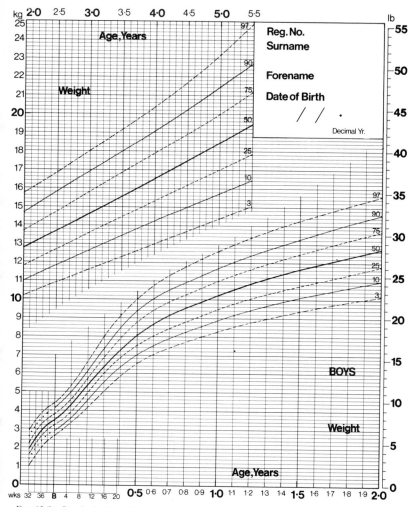

FIG. 65 (b). Standards for weight; ages 0–5: boys
(From Tanner and Whitehouse, 1973)

after birth, be plotted at age 6 months, or 26 postnatal weeks, but at 22 postnatal weeks. Failure to do this can result in considerable error in the first year after birth. Two scales are used in these charts, one from birth to 2 years, and a less expanded scale from 2 to 5 years (the top left sections). The standardizing data are exactly the same as in the 0–19 charts.

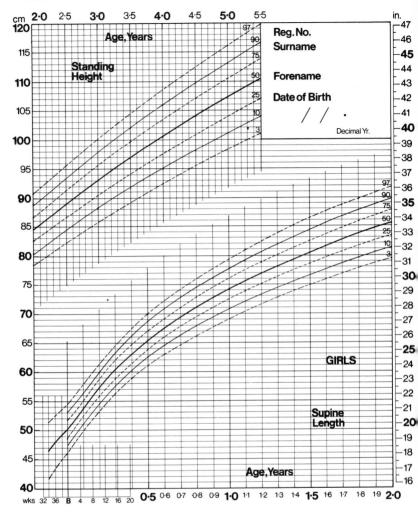

FIG. 66 (a). Standards for length–height; ages 0–5: girls
(From Tanner and Whitehouse, 1973)

United States Distance Charts

North American readers may wish to relate their height and weight to charts derived from U.S. data. For many years the pioneer data of Drs Howard Meredith at Iowa and Harold Stuart at Boston were used in American paediatric textbooks and found their way, perhaps unfortunately, halfway round the developing world. (Both

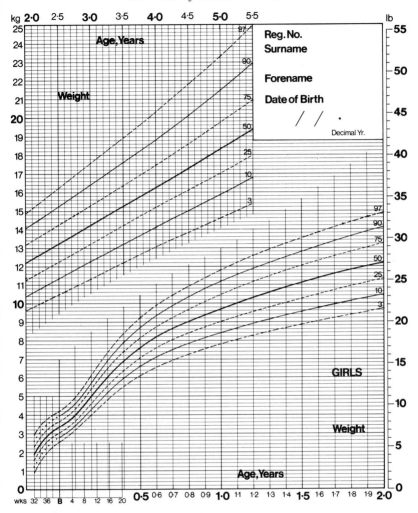

FIG. 66 (b). Standards for weight; ages 0–5: girls
(From Tanner and Whitehouse, 1973)

men were outstanding pioneers of American growth studies in the 1930s, the former a human auxologist and the greatest teacher of anthropometric technique of his generation, and the latter what would nowadays be called a Community Physician particularly concerned with the effects of disease and nutrition on the growing child. Their data, however, were derived from small and quite unrepresentative samples, and were satisfactory only in the days when there was nothing better.) Recently, a new set of charts has been issued by the National Center for Health Statistics (N.C.H.S.). These charts are based from birth to 3·0 years on the relatively limited and relatively selective data of the Fels Research Institute collected in the 1960–75 period. From 3·0 to 18·0 years, however,

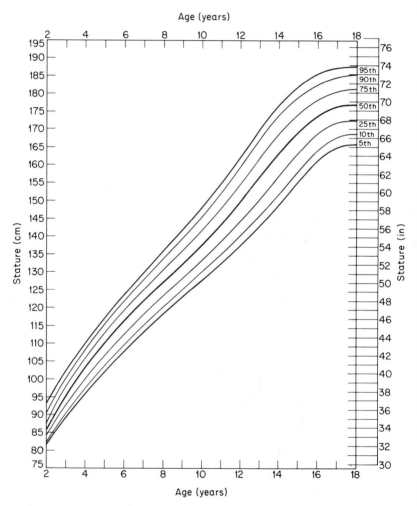

Fig. 67. Standards for height for U.S.A. boys
(Data from National Center for Health Statistics. Cross-sectional-type charts)

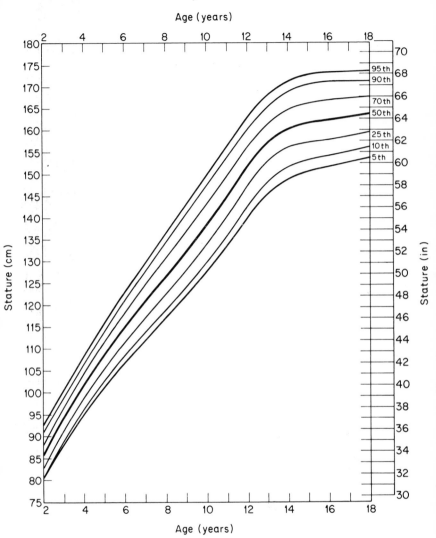

Fig. 68. Standards for height for U.S.A. girls
(Data from National Center for Health Statistics. Cross-sectional-type charts)

the data come from two specially mounted surveys (Health Examination Survey, 1962–70, and Health and Nutrition Examination Survey, 1971–74). These surveys aimed to sample the whole child population of the U.S.A. in a proportionate way, i.e. with numbers in each geographical area proportionate to the child

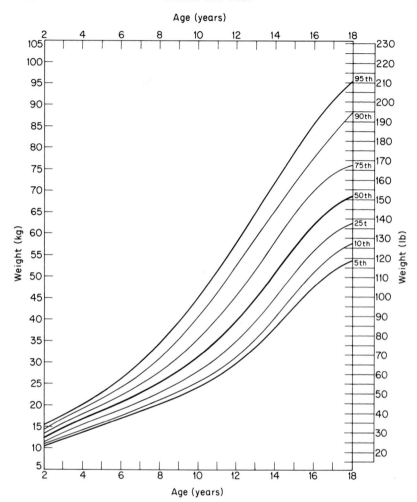

FIG. 69. Standards for weight for U.S.A. boys
(Data from National Center for Health Statistics. Cross-sectional-type charts)

population there. The resulting standards for height and weight are given in Figures 67 to 70, reproduced from the paper by Waterlow *et al.*, 1978. The 5th and 95th centiles are given rather than the 3rd and 97th, because the numbers varied from 300 to 600 children in each yearly age group and the errors of locating the outside centiles are large unless the sample is a very big one.

It must be remembered that these are *distance* standards founded

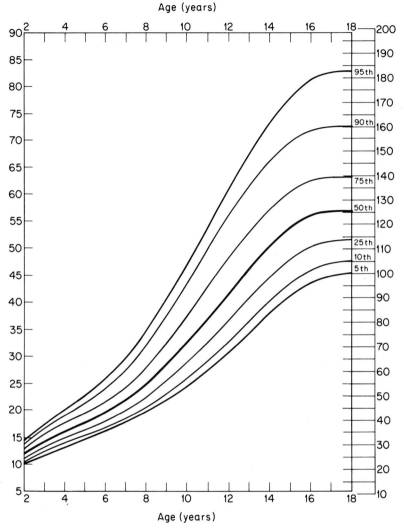

FIG. 70. Standards for weight for U.S.A. girls
(Data from National Center for Health Statistics. Cross-sectional-type charts)

on cross-sectional data reported simply, and not used, as in the British standards, to give limits around the true longitudinal-shaped curve. Thus they are more appropriate for use in comparing populations (for undernutrition, etc.) than for plotting individuals. Indeed they were designed specifically for the former purpose, and have been recommended for use as a W.H.O. reference standard for use in monitoring the nutrition of countries or subpopulations within countries.

Figures 71 and 72 show weight-for-height standards derived from the same source. These standards only apply to pre-pubescent children (of whatever age). This is because the weights of boys between the heights of 110 and 115 cm., say, might be different according to

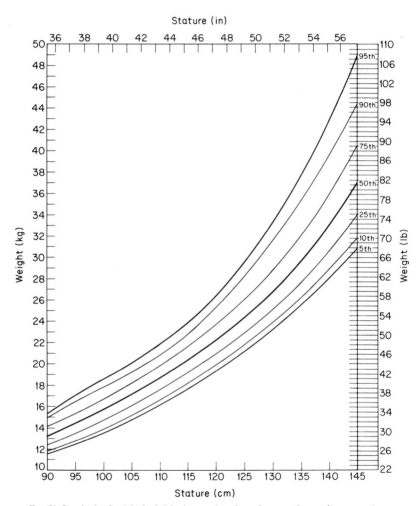

FIG. 71. Standards of weight for height, irrespective of age, for pre-pubescent boys, over the age of 1 year.

(Data from National Center for Health Statistics)

whether the boys in question were 5- or 6-year-olds of that height. If there was a difference, either in mean weight or in the range of weight, one standard of weight for height would not suffice for the two ages. In fact, weight for height is practically independent of age

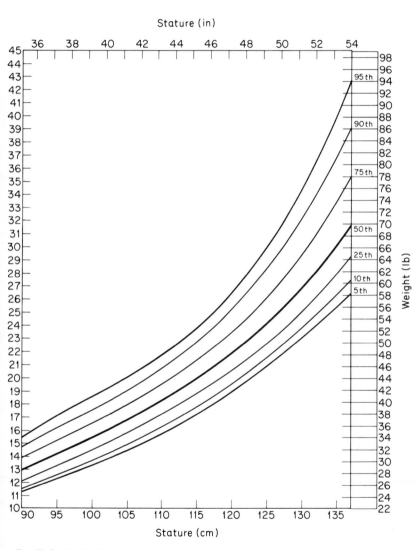

FIG. 72. Standards of weight for height, irrespective of age, for prepubescent girls, over the age of 1 year.

(National Center for Health Statistics data)

from age 1 to puberty; but when puberty begins this no longer holds. These standards can be used for all pre-pubescent individuals; it has to be remembered, however, that North American children are fatter than others, and that an optimal weight for height (optimal in the medical sense of healthiest) is probably well *below* the 50th centile.

Parent-allowed-for Charts

Figure 73 represents a different sort of chart. We have already seen how children resemble their parents, and clearly a small child with small parents represents a quite different situation from that of an equally small child with large parents. We can use the chart to allow for the heights of the parents and thus reach a more accurate assessment of whether the child has grown normally or not.

The chart was constructed with the help of the writer's colleagues R. H. Whitehouse and H. Goldstein. The data used were for parents and children from the whole range of the International Children's Centre Longitudinal Growth Studies, a group of linked studies in Brussels, London, Paris, Stockholm and Zurich, in each of which a cohort of children was followed from birth to maturity using the same techniques both for physical and psychological assessment. The chart is valid only for children aged 2·0 to 9·0 years; before 2, the relationship to parental height is not sufficiently developed, and after 9·0 the adolescent spurt makes things considerably more complicated.

Statistical analysis reveals that each parent on average is equally important in determining the child's height. Despite all popular belief to the contrary, there is no closer relation of father with son or mother with daughter than father with daughter and mother with son (see Chapter 9, p. 123). Consequently, we may use the simple average of parents' heights (uncorrected for sex) with which to compare the height of the child. This average is called mid-parent height and is placed on the horizontal scale of Figure 73 with indications of the 3rd, 50th and 97th centile of mid-parent height in the present-day U.K. population.

The form in which a child's height is related to mid-parent height can be expressed as what is called a regression equation. Thus the height H of a boy may be estimated as: $H = a + bX$ where X is mid-parent height and a and b are constants. Statistical technique enables us to calculate the likely range of variation around this estimate in a way exactly similar to the centiles we have already

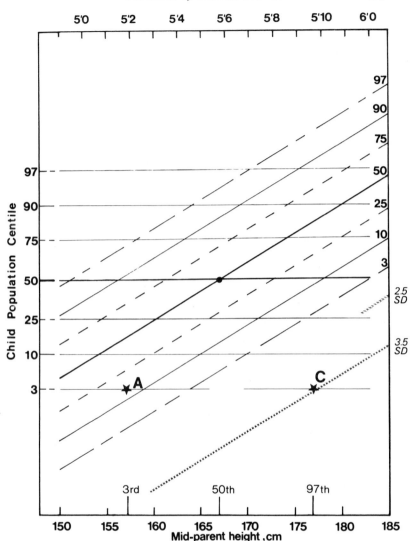

Fig. 73. Standards for height at ages 2·0 to 9·0, allowing for height of parents. For child population centile see Figures 55 or 56; centiles may be interpolated by eye. A and C are two children both at 3rd centile for population, with mid-parent centiles at 3rd and 97th.

met. A girl with parents of heights 160 cm. and 180 cm. has a height given by $H = a + 170b$, where a and b depend on the girl's exact age. If the girl is 8·0 years H is estimated as 127 cm. The 3rd and 97th centile limits to this estimate are 118 cm. and 136 cm. We now see how tall the girl actually is. If she is 116 cm. she is below the 3rd centile for her parents' height even though this value does not carry her below the 3rd centile for the population at large.

The chart is constructed so that the child's height centile has first to be looked up in the regular population charts of Figures 55 and 56. In this way Figure 73 may be used for both boys and girls. The regular population centiles constitute the vertical axis of the parent-allowed-for graphs; the exact centile can be estimated by interpolation with sufficient accuracy for our purpose of assessment. Suppose we have a boy at the 3rd centile in Figure 55. His parents are very small, with mid-parent height itself at the 3rd centile (157 cm.). Draw an imaginary line across the graph from the 3rd centile point on the vertical axis and another up the graph from the 3rd centile point on the horizontal axis; plot the point where the lines cross (i.e. A in the figure). Now consider A in relation to the system of diagonal lines, which represent the regression equation standard. The lines are marked in centiles (with indications of two lower ones at 2·5 and 3·5 standard deviations which may be ignored by the statistically innocent). The point A lies between the 10th and 25th lines, at about the 15th centile. The interpretation is straightforward. Though this boy is at the 3rd centile for the population of all boys of his age, he is at the 15th centile when his parents' heights are taken into account, and thus well within the limits of normal for his family. (Clearly, this interpretation, and indeed the whole use of this approach, is only valid if the parents have grown to their full height potential or somewhere near it. If the parents' heights have been curtailed by a poor environment the correlation between parents and children breaks down, as explained in Chapter 9.)

Consider a second boy, also at the 3rd centile for his population but with very large parents whose mid-parent centile is at the 97th (177 cm.). The resulting point is labelled C in the Figure. Here the situation is radically different; this boy is far below the 3rd centile for his family and in the clearly pathological range. He should be investigated (the child studied was one with asymptomatic coeliac disease). Rather than plot the intermediate point between A and C representing the 3rd centile boy with parents at the mid-parent 50th centile, let us leave readers who believe their statistical innocence has been lost, the question: 'Why is the intermediate point below, not on, the 3rd regression centile?'

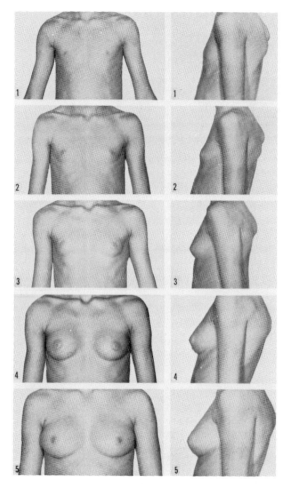

Fig. 74. Breast standards   (From Tanner, 1962)

Puberty Standards

Besides these standards for growth and growth velocity we can construct standards for assessing whether the various stages of puberty have been attained at the usual ages. The stages of breasts in girls, genitalia in boys and pubic hair in both sexes are illustrated in Figures 74 to 77. Extended descriptions need not be given here; suffice to say that breast stage 1 (B1) represents pre-adolescence, B2 the first appearance of what is called the breast bud, B3 a small

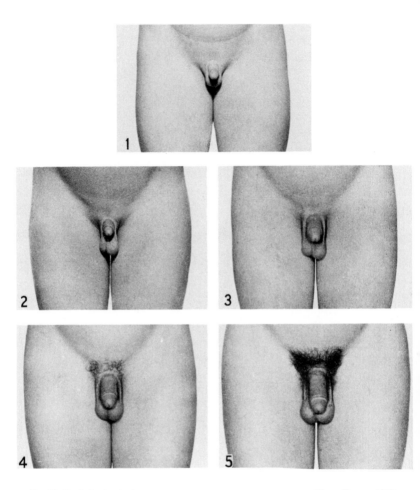

FIG. 75. Genitalia standards   (From Tanner, 1962)

pubescent breast, B4 a stage where the areola projects beyond the
contour of the breast (a stage skipped by some girls and persisted in
for several years by a few), and stage B5 the mature pre-pregnant
breast. Stages of pubic hair are the same for both sexes, with extension of hair up the abdominal mid-line in the male not counted in
the system.

The terminology is important; B2 signifies the very first moment
when B2 becomes visible, technically the transition from B1 to B2.
The stage B2 + lasts from B2 to B3. The girls' puberty-stages chart

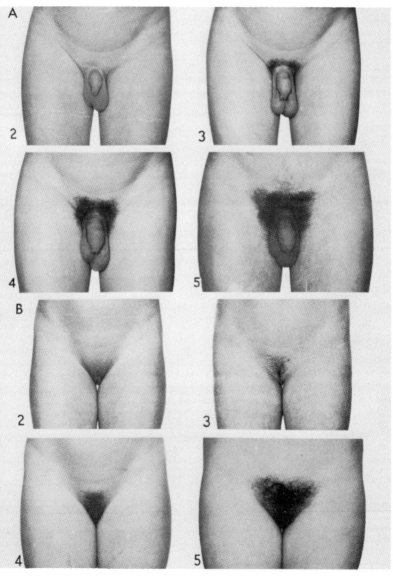

FIG. 76. Pubic hair standards: (*a*) boys; (*b*) girls   (From Tanner, 1962)

(Figure 77, lower) can now be readily understood. We cannot provide centiles for breast development at each age like those for height because we cannot measure the breast; we can only rate its appearance in terms of separate stages. Therefore we turn the standards round 90°, so to speak, and instead of asking 'Given this girl's

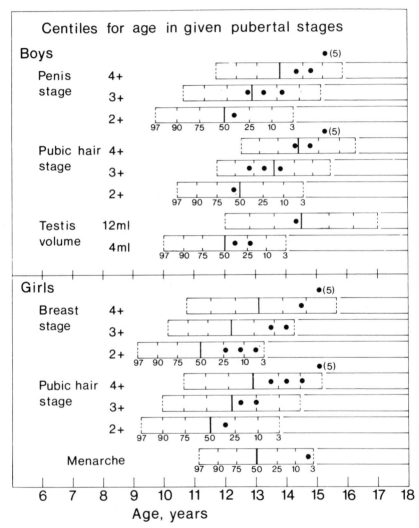

FIG. 77. Standards for puberty: (*a*) boys; (*b*) girls. One individual of each sex, serially followed through puberty, is plotted.
(From Tanner, 1962)

age, is her height within normal limits?' we ask 'Given this girl's breast appearance, is her *age* within normal limits?' The standards tell us that the average (50th centile) girl with the breast appearance B2 is aged 11·2 years; the early limit for such an appearance (quite arbitrarily labelled 97th centile rather than 3rd) is 9·2 years and the late limit is 13·2 years. Subsequent stages are plotted in the same way except that no age limits are possible for stage 5 since it persists indefinitely. The example shows ratings for a girl followed longitudinally. Her breasts developed considerably later than average, and her pubic hair slightly later; her menarche occurred at almost the late limit of the normal range.

The boys' chart (Figure 77, upper) is similar, except that the two lines at the bottom represent the attainment of two particular sizes of the testes as measured by palpation and comparison with the models of the Prader orchidometer (see Chapter 5, p. 60). Out of the progression of sizes these two are chosen because 4 ml. represents the first definitive enlargement at puberty and 12 ml. is a size reached at maturity by everyone (though some attain 20 or 25 ml.).

Standards for Bone Age

It is beyond the scope of this book to give standards for skeletal maturity or bone age. The left wrist and hand are x-rayed, and the appearance of the bones compare with standard appearances given in the Tanner–Whitehouse or Greulich–Pyle systems (see Chapter 6, pp. 79–81).

What emerges in the Tanner–Whitehouse system is a score for maturity. Maturity differs in an important way from a measurement such as stature, in that the normal growth process takes every individual from a common condition of being wholly immature to another of being wholly mature. Maturity can thus be measured on a scale of, say, 0 to 100, for everyone. An individual's score can then be compared with the centile distribution of scores of other individuals of the same age in just the same manner as height. We can then say whether skeletal maturity is advanced or delayed.

Alternatively, we can convert this maturity assessment into a bone age: the age at which the average child of the standard population has this particular maturity score. Thus if a boy has a maturity score of 60 units at age 12·0, one looks up a table and discovers that 60 is the score shown by the *average* boy at 10·0 years. The boy in question is therefore said to have a bone age of 10·0 'years', i.e. one 2 years retarded in relation to his chronological age.

The open-circle plot of Figure 55 (p. 169) can now be understood; and the diagnosis 'short stature due to growth delay' envisaged. This diagnosis is discussed further in the next chapter. At ages greater than 3·0 years the limits of normal bone-age delay and advance are about 2 years in either direction. Beyond this, pathology may be expected. The limits are narrower from 2 to 3 years and before age 2 skeletal maturity derived from the hand and wrist is inaccurate owing to lack of sufficient bones in the x-ray; radiographs of knee and ankle have to be used instead.

Prediction of Adult Height

A child's adult height may be predicted from the height of its parents, the most likely height centile being the average of the height centiles of mother and father (or, more precisely, the average of standard deviation-scores of mother and father). Such a prediction can be made even before the child is born. The error is fairly great (represented by a correlation coefficient of about 0·7) and a better prediction can be made from the child's own height, at least after the age of 2. From age 2 until the beginning of puberty the correlation between the child's present height and his adult height is between 0·80 and 0·85, which means that 95% of the time the error of prediction will be within the limits of about 7 cm. (3 ins.) either side of the predicted value. Though this may not sound very accurate, it does represent a considerable restriction compared with the whole population range which is ± 13 cm. about the average for men and ± 12 cm. for women. It is better, too, than prediction from parents' centiles, which gives an error of about ± 9 cm.

The simplest way to set forth these predictions is by giving a table of the average percentage of adult height reached at each age. Table 5 gives values for both the British standard longitudinal data

Table 5 (*opposite*)

a. Percentage of adult height reached at successive ages by average boy and girl in UK. Values during puberty (below lower horizontal lines) refer only to children with peak height velocity occurring at average time (12·0 and 14·0 years).
(Based on Tanner–Whitehouse longitudinal standards)

b. Percentage of adult height reached at successive years of age by boys and girls in the USA.
(Based on cross-sectional standards of N.C.H.S.)

TABLE 5

Age (yr)	[a] UK Boys	[b] USA Boys	[a] UK Girls	[b] USA Girls
0·08	30·9	30·8	32·7	32·6
0·25	34·7		36·4	
0·50	39·0	38·5	40·4	40·3
0·75	41·6		43·3	
1·00	43·7	43·0	45·7	45·3
1·25	45·4		47·8	
1·50	47·0	46·7	49·7	49·3
1·75	48·4		51·3	
2·00	49·8	49·4	52·8	52·7
2·00	49·2	48·8	52·1	52·7
2·5	51·6		54·8	
3·0	53·9	53·8	57·3	57·4
3·5	56·1		59·7	
4·0	58·2	58·1	61·9	62·0
4·5	60·1		64·0	
5·0	62·0	62·0	66·1	66·4
5·5	63·8		68·0	
6·0	65·6	65·7	69·9	70·0
6·5	67·3		71·8	
7·0	69·0	68·9	73·6	73·6
7·5	70·6		75·3	
8·0	72·2	71·7	77·1	77·2
8·5	73·8		78·8	
9·0	75·4	74·6	80·5	80·6
9·5	76·9		82·2	
10·0	78·3	77·6	83·8	84·5
10·5	79·8		85·5	
11·0	81·3	81·1	87·3	88·4
11·5	82·7		89·5	
12·0	84·1	84·6	92·2	91·1
12·5	85·5		94·7	
13·0	87·1	88·1	96·7	96·1
13·5	89·1		98·0	
14·0	92·0	92·1	98·9	98·0
14·5	94·6		99·5	
15·0	96·6	95·4	99·8	99·1
15·5	97·9		99·9	
16·0	98·8	98·1	100·0	99·2
16·5	99·4		—	
17·0	99·8	99·7	—	99·5
17·5	99·9		—	
18·0	100·0	100·0	—	100·0

and the American N.C.H.S. data. The table shows very well the advancement of girls as compared with boys and vindicates the old wives' tale that a boy doubles the length he has on his second birthday – but *girls* more nearly double the length they have at 18 months. The prediction should only be made for those ages placed between the horizontal lines, i.e. 2·0 to 10·0 years for boys and 2·0 to 8·0 years for girls. Beyond 10 in boys and 8 in girls the percentages apply only to those boys and girls who have their adolescent growth spurts at exactly the average time, with peak height velocity at 14·0 and 12·0 years respectively.

During adolescence these simple predictions become much less accurate because they fail to take into account whether the child is an early maturer and hence at a given age has little more growth to occur, or a late maturer with the full adolescent spurt yet to come. During this period it is essential to include in the prediction a measure of bone age. When this is done the prediction becomes quite accurate again. The limits within which 95% of predictions fall are, for boys, ± 7 cm. up to age 12, ± 6 cm. at ages 13 and 14, ± 5 cm. at age 15, and ± 4 cm. at age 16. For girls, the errors are approximately ± 6 cm. up to age 11, ± 5 cm. at age 12 if the girl has not yet had menarche but only ± 4 cm. if she has passed menarche, ± 4 cm. at 13 if premenarcheal and ± 3 cm. if postmenarcheal, and ± 2 cm. at age 14.

When bone age is used the prediction is made through regression equations specific for each age. As an example, consider a boy aged 11·0 years. His bone age on the Tanner–Whitehouse system ('R U S bone age', meaning bone age based on radius, ulna, metacarpals and phalanges only) is, say, 12·2 'years'. The equation for predicting his adult height is:

Adult height $= 1·16$ (present height)
$$ - 5·5 \text{ (chronological age)} - 1·6 \text{ (bone age)} + 89.$$

Tables of the regressions and full details of their use will be found in *Assessment of Skeletal Maturity and Prediction of Adult Height* (Tanner *et al.*, 1975). A small further correction may be made for parents' heights, but the exact extent, when the initial prediction is made in this manner, is not at present clear. Prediction may also be made by use of the Greulich–Pyle atlas of bone age, combined with tables provided by Dr Nancy Bayley, a psychologist by profession, whose papers on physical growth in the period 1930–60 rank with those of the two American 'immortals' Franz Boas (anthropologist, papers, 1880–1930) and Frank Shuttleworth (another psychologist, papers 1930–50). The more modern method is, however, somewhat the more accurate.

Prediction of adult height is not just an amusing pastime. In certain jobs height is a crucial qualification, and for many years the Department of Growth and Development at the Institute of Child Health has advised the Royal Ballet School as to the predicted height of children seeking to enter its school, usually at age 9 to 10. Ultimate entry to the Corps de Ballet is restricted to girls and boys within fairly narrow ranges of heights, so if the prediction clearly indicates that an entrant's chance of ending up within these limits is slender then both the child and the parents are told, and advised that though teaching, or other styles of dancing, may be possible, classical ballet is not. In this way, a terrible disappointment at the end of years of specialized training may be avoided.

Prediction of adult height is often most reassuring to tall girls believing, usually wrongly, that they are going to grow into 'giantesses', and to short boys, usually with considerable growth delay, who fear the opposite. The predictions do not apply, however, to children with endocrine or bone pathologies affecting growth.

CHAPTER 12
Disorders of Growth

This final chapter gives a brief description of the common disorders of growth, in terms shorn, so far as possible, of medical terminology. Growth disorders are at present claiming an increasing share of the family doctor's and paediatrician's attention as the infectious diseases of childhood are brought under control, and in the U.K. specialized growth-disorder clinics have been set up in numerous university medical centres, equipped with proper instruments for measuring children and assessing their maturity, and staffed by persons trained in auxological methods. Such clinics have particularly close working relationships with departments of paediatrics, endocrinology, medical genetics, orthopaedics and psychiatry.

Causes of Short Stature
Most of the children coming to such a clinic have short stature as their main complaint (we *never* use the word dwarfism: to many parents and children it has a curiously magical ring, derived from an amalgam of medieval folk-tales, grotesque gargoyles, Wagnerian Nibelungen and Scandinavian troll-kings. It is a term better totally avoided).

Short stature may be due to:

1. Normal genetic shortness
2. Growth delay
3. Chromosomal abnormalities
4. Intra-uterine maldevelopment
5. Endocrine gland disorders
6. Cartilage and bone disorders
7. Disorders of absorption of food
8. General diseases of the kidney, heart, etc.
9. Psychological causes

These will be discussed in turn.

1. Normal Genetic Shortness
Standards for children's height based on the whole population of children are only a first convenience. If a child is short, i.e., by

convention, below the 3rd centile, he should be assessed (if between ages 2·0 and 9·0 years) on the more accurate standards which allow for his parents' heights, as described in Chapter 11, p. 195. If he is within normal limits for his parents then he himself is normal, whatever the population position, unless, of course, one or other of his parents has a growth disorder. Note that both parents should be seen and actually measured. Estimates of height are notoriously inaccurate, particularly estimates by wives of their husbands' heights (they usually increase it; sometimes the reverse).

There is no effective treatment for genetic short stature (and persons who value the variability of our species may well add 'Long may it remain so'). So far as we know, families of short people secrete no less growth hormone than families of tall ones, and certainly the short cannot be made tall by giving them excess of growth hormone, which is simply without effect.

2. Growth Delay

This is a very common cause of attendance at a growth disorder clinic; thus every child attending has his bone age assessed. Delay may be manifest in early childhood as well as at the age when puberty should normally occur. Delay of up to 2 years is perfectly normal, and smaller numbers of normal children are delayed 3 or even 4 years. Frequently one or other of the parents, or an aunt or uncle or brother or sister, has also been delayed. The mother is asked about her age at menarche, which may well be 15 or 16 years instead of the more usual 12 to 14 years. Sometimes a history of delay in the father's puberty can be obtained (on occasion from his wife) but in general this is more difficult owing to the lack of a prominent landmark like menarche.

Growth delay is not in itself abnormal; puberty occurs in the end, and develops in every way quite normally. Thus boys and girls with this diagnosis can be absolutely reassured, with a full demonstration of facts and figures, charts and predictions. Children find the information that they are simply taking after their mother or father reassuring, if not particularly welcome. What reassurance does not do is remove the very real physical handicap associated with being small and still more with being pre-pubertal in a society of adolescents. The situation is often compounded or precipitated by a younger brother or sister who ploughs on ahead while the delayed child hangs about at the lowest growth velocity he or she has ever experienced (the immediately pre-adolescent velocity being so low that parents sometimes allege, mistakenly, that the child has actually stopped growing altogether for a year or more).

Parents, child and doctor may be tempted to give hormonal treatments to accelerate growth, particularly in boys. Testosterone is indeed effective for this; the problem is that it also accelerates

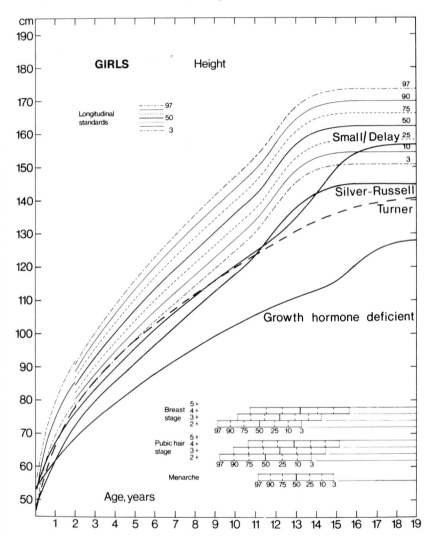

FIG. 78. Typical height curves of girls with growth delay (small/delay), Silver–Russell syndrome, Turner's syndrome and isolated growth-hormone deficiency
(From Tanner, Lejarraga and Cameron, 1975)

bone age, frequently to a greater relative degree than height. In this case the final height of the boy is diminished, usually too high a price to risk paying, particularly as growth-delayed children who come to the clinic tend also to be genetically somewhat small, presumably because these are the ones who feel the pinch most. There are a number of drugs similar to testosterone available whose actions are alleged to be less on bone age and sex characters than on the height spurt. Such claims at present should be treated with caution. If such a treatment is undertaken it should only be done by a physician with specialist training in auxological problems. Figure 78 shows a typical growth curve of a child with growth delay, compared with curves of children with other diagnoses.

3. Chromosome Abnormalities

The two most common chromosome abnormalities producing short stature are having 3 instead of 2 chromosomes No. 21, and having 1 instead of 2 X chromosomes in a girl. The former abnormality (see p. 55) produces Down's syndrome (also called mongolism, because of the appearance of the eyelid). The child is mentally handicapped as well as small. The appearance and behaviour makes diagnosis at an early age relatively easy; it is confirmed by examination of the karyotype (see Chapter 4, p. 52).

The lack of the second X chromosome produces Turner's syndrome (p. 54). The child is a girl of short stature, with or without some special physical peculiarities, such as puffy hands and feet soon after birth, webbing of the neck, broad chest, small nipples and characteristic face. Her bone age is delayed only slightly or not at all until the time puberty should occur. Because she lacks ovaries no pubertal development of the breasts takes place and there is no adolescent growth spurt. Sometimes pubic hair develops sparsely, sometimes not at all. The diagnosis is confirmed by a karyotype. No treatment can make good the lack of ovaries, and the news of life-long sterility has to be broken to parents and child. The breasts can be made to grow by giving oestrogen, which sometimes is accompanied by pubic hair growth as well. An artificial 'menstruation' can be produced by giving oestrogen for 3 weeks followed by a week without treatment or on a progesterone-like hormone. These girls commonly wish to menstruate like their friends, at least until the time when their friends begin to give it up. No treatment is effective in increasing height in Turner's syndrome to any marked degree though some male-sex-hormone-like preparations similar to those used in growth delay in boys are under test at the time of writing. The cause of the short stature remains entirely obscure. Growth hormone secretion is normal. The final height averages about 140 cm., though it depends on the height of the parents, since the parental correlation seems, curiously, to be maintained.

The above describes the situation in a full Turner's syndrome, karyotype written XO. Mosaics (see p. 54) also occur in which only a proportion of cells have this karyotype, others being the normal female XX. In such cases the symptoms are reduced and all gradations from normal to the complete Turner's syndrome are seen, both as regards sex development and stature. It follows that any girl with short stature of unknown cause should have a full karyotype done, with a substantial number of cells examined for their chromosome complement. Sometimes diagnosis may actually be impossible, since a minor XO cell line may produce some small effects and then die out, leaving only XX cells.

4. Intra-uterine Maldevelopment

Maldevelopment in utero may be due to a fault in the fertilized ovum, a fault in the placenta which limits the supply of foods and oxygen to the foetus, or to disease or starvation of the mother. Certain of the first and all of the second and third sorts of maldevelopment cause a restriction of growth and hence lead to small size at birth. This smallness of size may be known only through weight. Though length would often be more informative, it is unfortunately not routinely taken in clinics in the U.K. although accurate apparatus for measuring it now exists.

Standards for birthweight are available (see Bibliography); the centiles are different for girls and boys, and for first-born and later-born – and, of course, for each week of gestational age. We are concerned here only with small-for-dates babies, i.e. those who are below the centile limits for weight at their particular gestational age. Babies who are simply born early at the normal weight for that length of gestation seem mostly to catch up to perfectly normal height. Small-for-dates babies may be born at term, or may be born after a shorter gestational period.

The majority of such small-for-dates children, particularly those born at term, develop within the normal centiles, though they do not fulfil their genetic potential, and hence appear somewhat small for their parents' heights. Their average centile seems to be about the 30th. However, some fail to catch up in this way. They remain short and lacking in subcutaneous fat, and have a characteristic facial appearance. The face as a whole is triangular, with large eyes and a small lower jaw; the forehead is large and prominent in relation to the face, and the ears are set low in the head and tend to stick out. The bridge of the nose is usually depressed and the mouth turned down at the corners (giving rise to the expression 'shark-mouthed'). A proportion of these children have marked asymmetry of the limbs or body, one arm or leg being longer than the other, or one side of the face or chest more developed. The condition constitutes a specific syndrome, known by the name Silver–Russell after the two paediatricians who independently described it in the 1950s. Mental development is usually normal.

There seems to be at present no effective treatment for these children, who grow up quite normally in every way except for size. They are healthy and active and should not be regarded as delicate. In middle childhood they put on some subcutaneous fat, and their puberty occurs at the normal time and in the normal sequence. The syndrome hardly ever occurs twice in the same family, so the mother of such a child can be reassured about subsequent pregnancies. What the cause of the disorder is, we do not know. The prognosis may be affected by the treatment given low-birthweight babies nowadays, which consists of feeding them more intensively than before, and keeping them warmer.

Some of the children with relatively low birthweights and subsequent short stature have rarer and more specific syndromes of maldevelopment (such as de Lange's syndrome, Prader–Willi syndrome and the syndrome due to excessive intake of alcohol by the mother during pregnancy, for which see p. 47). The number of such syndromes is only rivalled by the paucity of children who suffer from each. Most are probably due to intrinsic faults in the fertilized ovum, and some are inherited. The subject is a highly specialized one, only adequately tackled by paediatricians and medical geneticists with a particular interest in it.

5. Endocrine Gland Disorders

The two chief endocrine disorders which cause short stature are growth hormone deficiency and thyroid deficiency. A third cause is disease of the adrenal or other glands, leading to precocious puberty; the child in this case grows fast and develops the changes of puberty pathologically early. At this time the child is very large for his or her age, but as growth stops early the final result is short stature.

Growth Hormone Deficiency. A chart of the growth of two brothers with deficiency of growth hormone has already been shown (Figure 52, p. 157). The deficiency does not affect intra-uterine growth, at least to an extent measurable by birthweight, but

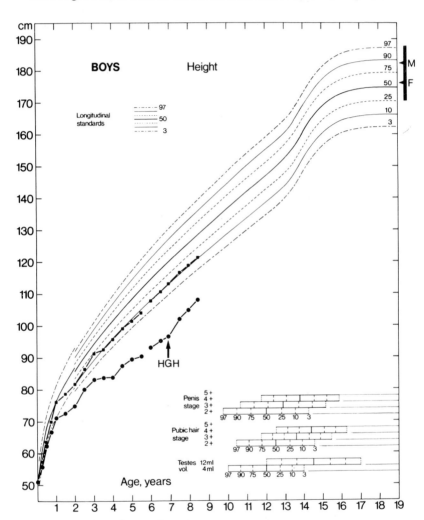

FIG. 79. Growth of two monozygotic twins, one with isolated growth-hormone deficiency. Measurements after break in lines made in Growth Disorder Clinic. Human growth-hormone treatment, begun at arrow, has continued.

from birth onwards growth is slower than normal. This is seen in Figure 79, which illustrates the height curves of two monozygotic twins, one with isolated growth-hormone deficiency. Their father was an engineer and their mother a teacher of physical education, and the children were both measured from birth onwards simply because of the interest of their being twins. The parents soon realized something was wrong, and eventually growth-hormone deficiency was recognized – much later than it should have been, for the whole subject was very new at that time. If all children had their length measured in pre-school clinics most cases of this disorder could be recognized by age 2 and proper and wholly effective treatment promptly instituted.

Most cases of growth-hormone deficiency are called 'idiopathic', meaning 'of unknown origin'. Others are associated with tumours of the central nervous system which destroy the part of the brain controlling the pituitary gland (see pp. 89–91), the pituitary itself, or both. Such tumours are nearly always benign, i.e. not cancerous, and the one which is by far the most common, craniopharyngioma, is not technically a brain tumour at all, but part of a piece of mouth epithelium pinched off and left behind during development. Craniopharyngiomas can be recognized in a skull x-ray or by computer-assisted tomography and most are readily removable. Occasionally thereafter they recur, but are removable again without great danger. The child with such a tumour occasionally comes to the physician with the sole complaint of short stature, but much more usually the symptoms are neurological and characteristic of a tumour in the head. After removal of the tumour the child may suffer from any combination of pituitary hormone deficiencies. Most, though not all, have growth-hormone deficiency. They are treated with human growth hormone, often in conjunction with thyroxine, cortisol (hydrocortisone), posterior pituitary hormone and, at the time of puberty, gonadotrophins, or sex hormones.

Idiopathic growth-hormone deficiency occurs, it seems, in about 1 in 10,000 births in the U.K. population at present. This may, however, change, for one of the chief clues to its cause is its association with breech birth, or the use of forceps in mothers who have already had one baby (and therefore seldom require forceps). Whether the breech birth contributes to its causation, or whether, for some obscure reason, lack of growth hormone is a factor contributing to the breech position, is unknown. But as many obstetricians are beginning to deliver by caesarian section babies who otherwise would be breech births we may soon know the answer to this.

There is evidently a hereditary predisposition, at least in some cases. About 3% of cases have brothers or sisters with the disorder, like the brothers illustrated in Figure 52 (p. 157). In a very few families one of the parents is affected. Curiously, the deficiency is four times as common in boys as in girls, for reasons quite unknown.

These children are simply small, with normal skeletal proportions, facial appearance and intelligence. They are usually fat (the condition diminishes on treatment) and they have a delayed bone age. The diagnosis is confirmed by a lack of growth-hormone production in response to a stimulation test, the stimulus being a sudden drop in blood sugar caused by an intravenous injection of insulin. The deficiency may be of growth hormone only, or other pituitary hormones may also be affected. In most cases in the U.K., growth hormone alone is involved; the next most common associated deficiences are those of TSH, the thyroid-stimulating hormone, and of FSH and LH, the gonadotrophins. ACTH, the adrenal-stimulating hormone, is much less frequently involved.

The treatment of growth-hormone deficiency has been a major success of paediatrics since the first patient was given human growth hormone by Raben in 1958 in Boston. Since the hormone is species-specific (see p. 92) human growth hormone extracted from corpses at autopsy has to be used. Fortunately adults' pituitaries contain as much growth hormone as children's, though what it does in adults we do not really know. The hormone is give by intramuscular injection 2 or 3 times a week for the whole duration of childhood and adolescence until growth stops naturally. Provided treatment has started at a reasonably early age (at least before age 6) the results are nearly always excellent (even at later ages results are sometimes spectacular, but not invariably). A catch-up occurs on initial treatment and thereafter normal growth is maintained. If TSH lack is also present, thyroxine is given, and if gonadotrophin deficiency declares itself at the time of puberty (which is very late in growth-hormone-deficient children, who have puberty at a normal bone age but a late chronological age), then sex hormones have also to be supplied. Very occasionally antibodies develop to the injected hormone and cancel out the effects of treatment; otherwise side-effects are practically unknown.

Hypothyroidism. Lack of thyroid gland secretion also stops growth occurring, for all cells seem to need a certain level of thyroid hormone in them in order to function properly. Hypothyroidism may start in utero, in which case the brain is affected and diagnosis and treatment directly after birth is a matter of urgency. The condition can be diagnosed by measuring the level of thyroid hormone in the blood, and recently in some areas this has been instituted as a routine screening-test for all newborns.

Short stature is more usually caused by the type of hypothyroidism which starts during childhood, often insidiously. Figure 80 shows an example. The catch-up on thyroid hormone here is marked and eventually complete, as is usually the case.

6. Cartilage and Bone Disorders

There are a large number of bone disorders which affect growth; they are mostly rare and many are inherited. The best known is *achondroplasia.*

The achondroplasiac is the classical dwarfed circus performer, though this is an image such societies as the U.K. Association for Research in Restricted Growth (ARRG) and the Little People of America would like to expunge. Achondroplasiacs have short upper arms and thighs, a normal length back, a large head, and a characteristic face, with depressed nasal bridge, small nose and large forehead. They usually have large muscles and are excellent athletes, whence their gravitation to circus work. They are of normal intelligence and health and usually seem to support their disorder with great cheerfulness and fortitude. Achondroplasia is due to the mutation of a single gene and is inherited as a dominant. This means that on average half the children of an achondroplasiac are affected. Most cases, however, are themselves the new mutation, without any family history; it seems that this particular gene is one of the most unstable in the human gene complex. There is no effective treatment at present.

A similar but less marked syndrome is *hypochondroplasia.* In this condition also only the limbs are short, and the legs are responsible for the short stature. However, the face is normal (though it has some tell-tale features for the experienced eye) and hypochondroplasiacs pass as very short normals. The disorder is also inherited as a dominant gene, but seems to be separate from the achondroplasia gene; at least the two conditions very seldom appear in the same family. In hypochondroplasia one of the parents frequently has the disorder. There is no absolutely certain way of clinching

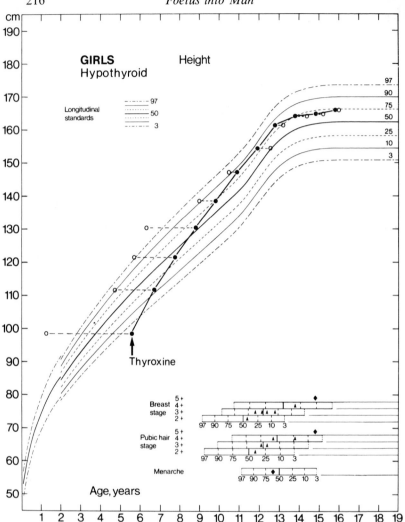

FIG. 80. Response of hypothyroid child treated with thyroxine. Solid circles, height for age; open circles, height for bone age
(From Tanner, Whitehouse, Marshall *et al.*, 1975)

what is sometimes a difficult diagnosis. The most helpful feature is comparison of sitting-height and leg-length centiles. While in achondroplasia the x-ray signs are marked and characteristic, in hypochondroplasia they are minimal and sometimes uncertain.

Other bone disorders are rarer; some affect limbs only, some the trunk only and some both. To date none are treatable.

For these disorders particularly, the ARRG extends extremely valuable support. It is a social association of people of short stature, together with parents of children of short stature. The members, who include doctors and surgeons with and without short stature, help one another with advice on all aspects of living, and educate the public into a better acceptance of the person handicapped in this way. Membership of the ARRG is usually the only thing a physician can offer parents of children with such bone disorders and it is frequently found to be of help.

7. Diseases of Absorption of Food

If there is prolonged malabsorption of food, growth is stunted, just as it is in starvation. All malabsorptive diseases may cause this if not treated successfully, but the most common one to appear in the growth disorder clinic is *coeliac disease*. Usually coeliac disease causes gastro-intestinal symptoms, but in a few children these are lacking and the disease is manifested solely in short stature, often but not always allied with an unusually protuberant abdomen. The diagnosis can only be certainly made by taking a biopsy of the small intestine through a tube that is swallowed. This sounds traumatic for a child, but nowadays in the skilled hands of the paediatric gastroenterologist it is not so. Coeliac disease is due to an abnormal reaction of cells of the gut to substances called glutens which occur in flour and other foodstuffs. The results of treatment by a gluten-free diet are so excellent that any child with short stature not otherwise diagnosed who is relatively thin should have a biopsy to exclude the possibility of coeliac disease.

8. General Diseases of the Kidney, Heart, etc.

A number of chronic general diseases cause short stature, mostly for reasons that are far from clear. Kidney disease is a prime example, and those heart and lung diseases which cause some degree of lack of oxygen to the tissues are others. Many of the heart lesions, however, do not stunt the child's growth at all.

9. Psychological Causes

It is now well established that psychosocial stress may cause short stature in certain children, who react to it by switching off their growth-hormone secretion. These are not starved children, and they are often fat, as in ordinary growth-hormone deficiency. There is a history of psychosocial stress; the child seems often to present an

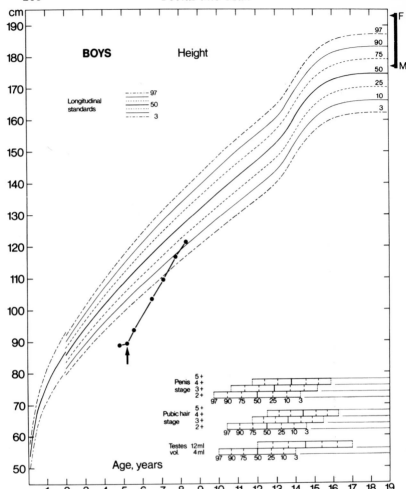

F_{IG}. 81. Height of child with psychosocial short stature. Note catch-up on removal, marked by arrow, from parental home.

older version of the battered baby situation. One child may be the scape-goat for the whole family, just as in the battered baby syndrome, with the other children growing quite normally.

In the classical case the child has a disorder of eating in that he eats voraciously at some times and not at all at others. He sleeps badly, gets up in the night and 'steals' food. Homes with children with psychosocial short stature nearly always have a lock on the refrigerator door. The child may eat the food set aside for the dog or cat, placing himself on the mat to do so; he rummages in dustbins and drinks water from the toilet. A stimulation test of growth-hormone secretion done in the home or immediately on arrival in the clinic shows inability to respond; the child is admitted to hospital and after a few days or weeks the response returns.

Once the diagnosis is fully established there is little to do but separate the child from the inimical family atmosphere (if he is sent to a school for 'delicate' children 'because he is so small' the family morale may be saved and agreement secured). Away from home the catch-up is usually spectacular and fully equal to that of an idiopathically growth-hormone-deficient child given growth hormone. An example is shown in Figure 81.

CODA

This chapter brings us to the end of our brief conspectus of human growth and its disorders. Provided always that parental size is known, growth emerges as the prime measure of a child's physical and mental health. The study of growth emerges also as a powerful tool for monitoring the health and nutrition of populations, especially in ecological and economic circumstances that are sub-optimal. It is equally powerful for studying the effect of political organization upon the relative welfare of the various social, cultural and ethnic groups which make up a modern state. Thus the study of growth has a very direct bearing upon human welfare. At the same time it gives us valuable lessons on the way in which our biological heritage and our technological culture interact. It warns us that all too soon we shall be technically capable of creating monstrously specialized or monstrously similar children. It points to our need to be reconciled with our origins, to see ourselves once again as a part of the natural order; not foetus into angel, nor foetus into monster, but foetus into Man.

Bibliography

Chapter 1

Further Reading

Falkner, F. and Tanner, J. M., eds. (1978), *Human Growth*, 3 vols., New York and London: Plenum.
The first comprehensive multi-author handbook on human growth. Chapters 6, 7 and 8, by M. J. R. Healy, H. Goldstein and E. Marubini respectively, are a particularly valuable introduction to biometrical methods in growth work, curve fitting and sampling for surveys. An elementary knowledge of statistics is assumed

Marshall, W. A. (1977), *Human Growth and its Disorders*, London and New York: Academic Press
A textbook for medical graduates, which deals simply but comprehensively with human growth, chiefly from a medical point of view

Tanner, J. M. (1962), *Growth at Adolescence*, 2nd edn, Oxford: Blackwell Scientific Publns
A textbook treatment of human growth, which, despite the title, covers the whole postnatal period in some detail. Suitable for readers with biological or medical background

Werff ten Bosch, J. J. van der and Haark, A., eds. (1966), *Somatic Growth of the Child*, Leiden: Stenfert Kroese
The proceedings of a Boerhaave course for postgraduate medical teaching at Leiden University. Very wide-ranging, covering all the topics in this book, with excellent speakers

References

Israelsohn, W. J. (1960), 'Description and modes of analysis of human growth', in *Human Growth*, ed. Tanner, J. M., Oxford: Pergamon, *Symp. Soc. Study Hum. Biol.*, **3**, 21–41

Tanner, J. M., Whitehouse, R. H. and Takaishi, M. (1966), 'Standards from birth to maturity for height, weight, height velocity and weight velocity; British children, 1965', *Arch. Dis. in Child.*, **41**, 454–71; 613–35

Tanner, J. M. and Whitehouse, R. H. (1975), 'Revised standards for triceps and subscapular skinfolds in British children', *Archives of Disease in Childhood*, **50**, 142–5.

Tanner, J. M. and Whitehouse, R. H. (1976), 'Clinical longitudinal standards for height, weight, height velocity, weight velocity and the stages of puberty', *Archives of Disease in Childhood*, **51**, 170–9.

Johnston, F. E., Hamill, P. V. V. and Lemeshow, S. (1974), 'Skinfold thicknesses in a national probability sample of U.S. males and females aged 6 through 17 years', *American Journal of Physical Anthropology*, **40**, 321–4 (also D.H.E.W. Publns Nos. HSM 73–1602 and HRA 74–1614)

Chapter 2

Further Reading

Goss, R. J. (1966), 'Hypertrophy versus hyperplasia', *Science*, **153**, 1615–20
A good short introduction to cellular growth
Goss, R. (1978), 'Adaptive mechanisms of growth control', in *Human Growth*, Vol. 1, eds. Falkner, F. and Tanner, J. M., New York: Plenum
A general and very readable account of tissue growth, with a description of experiments demonstrating some of the mechanisms of its control
Jacobson, M. (1974), 'Differentiation and growth of nerve cells', in *Differentiation and Growth of Cells in Vertebrate Tissues*, ed. Goldspink, G., London: Chapman & Hall
A clear and excellent résumé of the subject, suitable, however, only for readers with some knowledge of biology
Winick, M. (1971), 'Cellular growth during early malnutrition', *Pediatrics*, **47**, 969–78
A general account of the author's work on undernutrition and the brain in animals. Defends theory that the more rapid the rate of cell division at the time of starvation the more lasting the effect
Widdowson, E. M. (1970), 'Harmony of growth', *Lancet*, **1**, 901–5
A lecture, to a general audience, which provides a good starting-point for reading about comparative growth in the rat, pig, man, whale and other mammals. Emphasis is on cellular and tissue growth

References

Brasel, J. A. (1974), 'Cellular changes in intra-uterine malnutrition', in *Nutrition and Fetal Development*, ed. Winick, M., New York: Wiley
Brook, C. G. D. (1972), 'Evidence for a sensitive period in adipose-cell replication in man', *Lancet*, **2**, 625–7
Cheek, D. B. (1968), 'Muscle cell growth in normal children', in *Human Growth*, ed. Cheek, D. B., Philadelphia: Lea & Febiger
Goldspink, G. (1974), 'Development of muscle', in *Differentiation and Growth of Cells in Vertebrate Tissues*, ed. Goldspink, G., London: Chapman & Hall

Jacobson, M. (1970), *Developmental Neurobiology*, New York: Holt, Rhinehart & Winston

Pritchard, J. J. (1974), 'Growth and differentiation of bone and connective tissue', in *Differentiation and Growth of Cells in Vertebrate Tissues*, ed. Goldspink, G., London: Chapman & Hall

Steffe, W. P., Goldsmith, R. S., Pencharz, P. B., Scrimshaw, N. S. and Young, V. R. (1976), 'Dietary protein intake and dynamic aspects of whole body nitrogen metabolism in adult humans', *Metabolism*, **25**, 281–97

Waterlow, J. C. (1968), 'Observations on the mechanisms of adaptation to low protein intake', *Lancet*, **2**, 1091–4

Waterlow, J. C. (1975), 'Protein turnover in the whole body', *Nature*, **253**, 157

Chapter 3

Further Reading

Alberman, E. and Creasy, M. R. (1975), 'Factors affecting chromosome abnormalities in human conceptions', in *Chromosome Variations in Human Evolution*, ed. Boyce, A. J., *Symposia of the Society for the Study of Human Biology*, **14**, 83–96
A review of foetal mortality due to chromosome abnormalities

Campbell, S. (1976), 'The antenatal assessment of fetal growth and development: the contribution of ultrasonic measurement', in *The Biology of Fetal Growth*, eds. Roberts, D. F. and Thomson, A. M., *Symposium of the Social Study of Human Biology*, **15**, 15–38
An authoritative account of ultrasonic measurement of the foetus by an outstanding pioneer in that field

McLaren, A. (1968), 'Life before birth: the first few days', *Proceedings of the Royal Institution*, **42**, 153–69
A Royal Institution lecture, before a lay audience. Discusses embryonic development, differentiation in the embryo, and the techniques and results of egg transfer from one mother to another, in which subject the author was a pioneer

Neilson, L., Ingelman-Sunberg, A. and Winsen, C. (1967), *The Everyday Miracle: A Child is Born*, London: Lane, Penguin Press
A popular account of the growth of a child from conception to birth, distinguished by Neilson's beautiful pictures of human embryos and foetuses, which allow the reader to visualize exactly the changes of form, the situation in the uterus and the development of head, hands, feet and face

Tanner, J. M. (1974), 'Variability of growth and maturity in new-born infants', in *The Effect of the Infant on its Caregiver*, eds. Lewis, M. and Rosenblum, L. A., New York: Wiley
An account which stresses the extent of individual variability in size and maturity, even at birth

Walton, A. and Hamond, J. (1938), 'The Maternal effects on growth and conformation in Shire horse–Shetland pony crosses', *Proceedings of the Royal Society B*, **125**, 311–35
The classic experiment of crossing Shire with Shetland horses, to observe the size of the foal at birth and its subsequent growth

References

Birkbeck, J. A. (1976), 'Metrical growth and skeletal development of the human fetus', in *The Biology of Fetal Growth*, eds. Roberts, D. F. and Thomson, A. M., *Symposium of the Social Study of Human Biology*, **15**, 39–68

Brandt, I. (1976), 'Dynamics of head circumference growth before and after term', in *The Biology of Fetal Growth*, eds. Roberts, D. F. and Thomson, A. M., *Symp. Soc. Study Hum. Biol.*, **15**, 109–36

Brenner, W. E., Edelman, D. A. and Hendricks, C. H. (1976), 'A standard of fetal growth for the United States of America', *American Journal of Obstetrics and Gynecology*, **126**, 555–64

Butler, N. R., Goldstein, H. and Ross, E. M. (1972), 'Cigarette smoking in pregnancy: its influence on birth weight and perinatal mortality', *British Medical Journal*, **1**, 127

Campbell, S. and Wilkin, D. (1975), 'Ultrasonic measurement of fetal abdomen circumference in the estimation of fetal weight', *British Journal of Obstetrics and Gynaecology*, **82**, 689–97

Dargassies, S. St-A. (1966), 'Neurological maturation of the premature infant of 28 to 41 weeks gestational age', in *Human Development*, ed. Falkner, F., London: Saunders

Dreyfus-Brisac, C. (1975), 'Neurophysiological studies in human premature and full-term newborns', *Biological Psychiatry*, **10**, 485–96

Drillien, C. M. (1964), *The Growth and Development of the Prematurely Born Infant*, Edinburgh: Livingstone

Drumm, J. E., Clinch, J. and MacKenzie, G. (1976), 'The ultrasonic measurement of fetal crown–rump length as a method of assessing gestational age', *British Journal of Obstetrics and Gynaecology*, **83**, 417–21

Fancourt, R., Campbell, S., Harvey, D. and Norman, A. P. (1976), 'Follow-up study of small-for-dates babies', *British Medical Journal*, **1**, 1435–7

Jakobovits, A., Iffy, L., Wingate, M. B., Slate, W. G., Chatterton, R. T. and Kerner, P. (1972), 'The rate of early fetal growth in the human subject', *Acta anatomica*, **83**, 50–59

Knobloch, H. and Pasamanick, B. (1962), 'Mental subnormality', *New England Journal of Medicine*, **266**, 1045–51; 1092–7; 1155–61

Lechtig, A., Delgado, H., Lasky, R. E., Klein, R. E., Engles, P. L., Yarbrough, C. and Habicht, J. P. (1975), 'Maternal nutrition and fetal growth in developing societies', *American Journal of Diseases of Children*, **129**, 434–7

Moosa, A. and Dubowitz, V. (1971), 'Postnatal maturation of peripheral nerves in pre-term and full-term infants', *Journal of Paediatrics*, **79**, 915–22

Jones, K. L. and Smith, D. W. (1975), 'The fetal alcohol syndrome', *Teratology*, **12**, 1–10

Smith, D. W. (1977), *Growth and its Disorders*, Philadelphia: Saunders

Southgate, D. A. T. and Hay, E. N. (1975), 'Chemical and biochemical development of the fetus', in *The Biology of Fetal Growth*, eds. Roberts, D. F. and Thomson, A. M., *Symposium of the Social Study of Human Biology*, **15**, 195–209

Tanner, J. M. (1963), 'The regulation of human growth', *Child Development*, **34**, 817–47

Tanner, J. M. and Thomson, A. M. (1970), 'Standards for birthweight at gestation periods from 32 to 42 weeks, allowing for maternal height and weight', *Archives of Disease in Childhood*, **45**, 566–9

Usher, R. H. and McLean, F. H. (1974), 'Normal fetal growth and the significance of fetal growth retardation', in *Scientific Foundations of Paediatrics*, eds. Davis, J. A. and Dobbing, J., London: Heinemann

Chapter 4

Further Reading

McEwen, B. S. (1976), 'Interactions between hormones and nerve tissues', *Scientific American*, **235**, 48–58
An up-to-date account of the way in which sex hormones cause sex differentiation of the brain. Also discusses oestradiol and corticosterone receptors in brain. Very well illustrated

Money, J. and Ehrhardt, A. A. (1972), *Man and Woman: Boy and Girl*, Baltimore: Johns Hopkins University Press
Already a classic: written from the viewpoint of psychologists dealing with patients with intersexual and gender identity problems. The early chapters discuss sex differentiation of body structure and the later ones differentiation of gender identity and gender role. One of the very few books in which a satisfactory balance is held between the endocrinological and psychosocial aspects of sexual differentiation

Ratcliffe, S. C. (1975), 'Postnatal chromosomal abnormalities', in *Chromosome Variations in Human Evolution*, ed. Boyce, A. J., *Symposium of the Social Study of Human Biology*, **14**, 97–115
Gives worldwide data on incidence of chromosomal abnormalities at birth; some 43,000 presumed normal newborns had had karyotypes done at this time of reporting

References

Alberman, E. and Creasy, M. R. (1975), 'Factors affecting chromosome abnormalities in human conceptions', in *Chromosome Variations in Human Evolution*, ed. Boyce, A. J., *Symposium of the Social Study of Human Biology*, **14**, 83–95

Eayrs, J. T. (1971), 'Thyroid and developing brain: anatomical and behavioural effects', in *Hormones in Development*, eds. Hambrugh, M. and Barrington, E. J. W., New York: Appleton–Century

Feldman, K. W. and Smith, D. W. (1975), 'Fetal phallic growth and penile standards for newborn male infants', *Journal of Paediatrics*, **86**, 395–8

Josso, N. (1974), 'Mullerian inhibiting activity of human fetal testicular tissue deprived of germ cells by in vitro irradiation', *Pediatric Research*, **8**, 755–8

Niemi, M., Ikonen, M. and Hervonen, A. (1967), 'Histochemistry and fine structure of the interstitial tissue in the human foetal testes', in *Endocrinology of the Testis*, eds. Wolstenholm, G. E. W. and O'Connor, M., London: Churchill

Reyes, F. I., Winter, J. S. D. and Faiman, C. (1973), 'Studies on human sexual development. 1. Fetal gonadal and adrenal sex steroids', *Journal of Clinical Endocrinology and Metabolism*, **37**, 74–8

Chapter 5

Further Reading

Marshall, W. A. and Tanner, J. M. (1969), 'Variations in the pattern of pubertal changes in girls', *Archives of Disease in Childhood*, **44**, 291–303

Marshall, W. A. and Tanner, J. M. (1970), 'Variations in the pattern of pubertal changes in boys', *Archives of Disease in Childhood*, **45**, 13–23

Two papers describing pubertal variations between individuals, based on the largest number of children yet followed longitudinally and in detail throughout puberty

Tanner, J. M. (1962), *Growth at Adolescence*, 2nd edn, Oxford: Blackwell Scientific Publns

Covers pubertal changes in some detail, with extensive bibliography

Tanner, J. M. (1975), 'Growth and endocrinology of the adolescent', in *Endocrine and Genetic Diseases of Childhood*, 2nd edn, ed. Gardner, L., Philadelphia and London: Saunders

An up-to-date account of pubertal changes, and their endocrinological causes, suitable for biologists as well as doctors

Taranger, J., Engström, I., Lichtenstein, H. and Svennberg-Redegren, I. (1976), 'The somatic development of children in a Swedish urban community, VI, Somatic pubertal development', *Acta paediatrica scandinavica,* supplement, **258**, 121–35

A report of the Stockholm longitudinal growth study (parallel with studies in Brussels, London, Paris and Zurich) giving details in particular of ages of transition from each stage of puberty to the next

References

Jones, H. E. (1949), *Motor Performance and Growth. A Developmental Study of Static Dynamomometric Strength*, Berkeley: University of California Press

Marshall, W. A. and Tanner, J. M. (1974), 'Puberty', in *Scientific Foundations of Paediatrics*, pp. 124–52, eds. Davis, J. A. and Dobbing, J., London: Heinemann

Stolz, H. R. and Stolz, L. M. (1951), *Somatic Development of Adolescent Boys. A Study of the Growth of Boys during the Second Decade of Life*, New York: Macmillan

Tanner, J. M., Whitehouse, R. H., Marubini, E. and Resele, L. (1976), 'The adolescent growth spurt of boys and girls of the Harpenden Growth Study', *Annals of Human Biology*, **3**, 109–26

van Wieringen, J. C., Wafelbakken, F., Verbrugge, H. P. and de

Haas, J. H. (1971), *Growth Diagrams 1965, Netherlands*, Leiden: Institute of Preventive Medicine

Zachmann, M., Prader, A., Kind, H. P., Haflinger, H. and Budliger, H. (1974), 'Testicular volume during adolescence', *Helvetica paediatrica acta*, **29**, 61–72

Chapter 6

Further Reading

Acheson, R. M. (1966), 'Maturation of the skeleton', in *Human Development*, ed. Falkner, F., Philadelphia: Saunders

An excellent introduction to the assessment of skeletal maturity by a pioneer of the scoring method

Marshall, W. A. and Limongi, Y. (1976), 'Skeletal maturity and the prediction of age at menarche', *Annals of Human Biology*, **3**, 235–43

An important paper, in method as well as result. Requires only elementary knowledge of statistics, none of biology

Tanner, J. M. (1966), 'Galtonian eugenics and the study of growth. The relation of body size, intelligence test score and social circumstances in children and adults', *Eugenics Review*, **58**, 122–35, reprinted in *Trends and Issues in Developmental Psychology*, eds. Mussen, P. H., Largen, J. and Covington, M., New York: Holt, 1969

A review, citing the historical background as well as modern studies

Tanner, J. M., Whitehouse, R. H., Marshall, W. A., Healy, M. J. R. and Goldstein, H. (1975), *Assessment of Skeletal Maturity and Prediction of Adult Height*, London: Academic Press

A short monograph describing the Tanner–Whitehouse system of bone age and giving predictions of adult height based upon it. The early chapters discuss the general context of developmental age assessment

Tanner, J. M. (1978), *Education and Physical Growth*, 2nd edn, London: University of London Press.

A textbook for teachers. Contains more detail on the matters discussed in this chapter and Chapter 9, together with implications for education

References

Demerjian, A., Goldstein, H. and Tanner, J. M. (1973), 'A new system of dental age assessment', *Human Biology*, **45**, 211–27

Douglas, J. W. B. and Ross, J. M. (1964), 'Age of puberty related to educational ability, attainment and school leaving age', *Journal of Child Psychology and Psychiatry*, **5**, 185–96

Greulich, W. W. and Pyle, S. I. (1959), *Radiographic Atlas of*

   Skeletal Development of the Hand and Wrist, 2nd edn, Stanford,
   California: Stanford University Press

Marshall, W. A. (1974), 'Interrelationships of skeletal maturation,
   sexual development and somatic growth in man', *Annals of
   Human Biology,* **1**, 29–40

Mussen, P. H. and Jones, M. C. (1957), 'Self-conceptions, motivations and interpersonal attitudes of late- and early-maturing
   boys', *Child Development,* **28**, 243–56

Mussen, P. H. and Jones, M. C. (1958), 'The behaviour inferred
   motivations of late- and early-maturing boys', *Child
   Development,* **29**, 61–7

Roche, A. F., Wainer, H. and Thiessen, D. (1975), *Skeletal
   Maturity: The knee as a biological indicator,* New York: Plenum

Taranger, J. and Karlberg, P. (1976), 'The somatic development of
   children in a Swedish urban community, VII, Graphic analysis of
   biological maturation by means of maturograms', *Act paediatrica
   scandinavica,* supplement, **258**, 136–46

Chapter 7

Further Reading

Catt, K. J. (1971), *The ABC of Endocrinology,* London: Lancet
A standard textbook of endocrinology written in a straightforward and
simple style. Deals more with adults than children, however. Assumes
elementary knowledge of biochemistry

Grumbach, M. M., Grave, G. D. and Mayer, F. E. eds, (1974),
   Control of the Onset of Puberty, New York: Wiley
A collection of seventeen papers given at a conference, dealing with
animals as well as man. Useful only for readers with a biological or medical
background

McEwen, B. S. (1976), 'Interactions between hormones and nerve
   tissue', *Scientific American,* **235**, 48–58
An account of how steroid hormones act on cells, particularly those of the
brain. Much clearly-presented information, including experimental findings

O'Mally, B. W. and Schrader, W. T. (1976), 'The receptors of
   steroid hormones', *Scientific American,* **234**, 32–43
A beautifully clear and detailed account of how cytoplasmic hormone
receptors work, by two of the chief investigators in this field. Strongly
recommended to readers with some knowledge of biology

Tanner, J. M. (1972), 'Human growth hormone', *Nature,* **237**,
   431–7
A review article which assumes no special biological or medical knowledge.
Now a little out of date, but could serve as primary background for readers
interested in study of the original literature

References

Frisch, R. E. (1974), 'A method of prediction of age of menarche from height and weight at ages 9 through 13 years', *Pediatrics*, **53**, 384–90

Prader, A., Tanner, J. M. and von Harnack, G. A. (1963), 'Catch-up growth following illness or starvation', *Journal of Paediatrics*, **62**, 646–59

Chapter 8

Further Reading

Brandt, I. (1976), 'Dynamics of head circumference growth before and after term', in *The Biology of Humal Fetal Growth*, eds. Roberts, D. F. and Thomson, A. M., *Sym. Soc. Study Hum. Biol.*, **15**, 109–36

A review of the growth of head circumference, incorporating the results of the author's own very thorough longitudinal study of pre-term infants. Highlights difficulties of obtaining accurate information on head and brain growth

Changeux, J.-P. and Dauchin, A. (1976), 'Selective stabilisation of developing synapses as a mechanism for the specification of neuronal networks', *Nature,* **264**, 705–12

An excellent review, though suitable only for readers with a knowledge of physiology and biochemistry

Gaze, R. M. and Keating, M. J. eds., (1974), 'Development and regeneration in the nervous system', *British Medical Bulletin*, **30**, 105–89

Fifteen essays ranging from the basis of neuronal specificity to the development of infant behaviour. Biological or medical knowledge is assumed, but no special acquaintance with the nervous system

Jacobson, M. and Hunt, R. K. (1973), 'The origins of nerve-cell specificity', *Scientific American*, **228**, 26–35

Not an easy article, but gives a real insight into the painstaking world of experimental embryology. The authors describe their experiments on the mechanism by which nerve cells from the retina achieve the exact point-to-point projection on the brain centres subserving vision. A feel for three-dimensional structures is helpful

Llinas, R. R. (1975), 'The cortex of the cerebellum', *Scientific American*, **232**, 56–71

A good introduction to the relation between structure and function in the central nervous system. The cerebellum is better understood than other parts of the brain, and its relatively simple structure is beautifully illustrated in this article. No knowledge of physiology is assumed; straightforward to anyone used to wiring diagrams or blue-prints

Lynch, G. and Gall, C. (1978), 'Organization and reorganization in the central nervous system: evolving concepts of brain plasticity', in *Human Growth*, Vol. 3, eds. Falkner, F. and Tanner, J. M., New York: Plenum

A review summarizing the evidence that the brain and in particular the immature brain, reacts to an excess or a deprivation of particular stimuli by altering the size and structure of neurons and neuroglia.

Mark, R. (1974), *Memory and Nerve Cell Connections*, Oxford: Clarendon

A most stimulating introduction to the nervous system from the point of view of its development and organization. Though authoritative, it is not overburdened with technology, and can be understood by any reader who has an elementary idea of electronics or biology. For the more advanced, still an excellent introduction to neurophysiology

Sidman, R. L. and Rakic, P. (1974), 'Neuronal migrations in human brain development', *Modern Problems in Paediatrics*, **13**, 13–43

A highly technical account, but with beautiful reconstructions, and full of ideas

References

Blakemore, C. (1976), 'The conditions required for the maintenance of binocularity in the kitten's visual cortex', *Journal of Physiology*, **261**, 423–44

Burn, J., Birkbeck, J. A. and Roberts, D. F. (1975), 'Early fetal brain growth', *Human Biology*, **47**, 511–22

Campbell, S. and Newman, G. B. (1971), 'Growth of the fetal biparietal diameter during normal pregnancy', *Journal of Obstetrics and Gynaecology of the British Commonwealth*, **78**, 513–19

Cheek, D. B. (1975), *Fetal and Postnatal Cellular Growth*, New York: Wiley

Cheek, D. B., Holt, A. B., London, W. T., Ellenberg, J. H., Hill, D. E. and Sever, J. L. (1976), 'Nutritional studies in the pregnant rhesus monkey: the effect of protein-calorie or protein deprivation on growth of the fetal brain', *American Journal of Clinical Nutrition*, **29**, 1149–57

Conel, J. L. (1939–67), *The Postnatal Development of the Human Cerebral Cortex*, Vols. 1–8, Cambridge: Harvard U. P.

Cragg, B. G. (1968), 'Are there structural alterations in synapses related to functioning?', *Proceedings of the Royal Society B*, **171**, 319–23

Dobbing, J. (1974), 'Later development of the brain and its vulnerability', in *Scientific Basis of Paediatrics*, eds. Davis, J. A. and Dobbing, J., London: Heinemann

Howard, E., Granoff, D. and Bujnovsky, P. (1969), 'DNA, RNA and cholesterol increases in cerebrum and cerebellum during development of human fetuses', *Brain Research*, **14**, 697–706

Hubel, D. H. and Wiesel, T. N. (1963), 'Receptive fields of cells in striate cortex of very young, visually inexperienced kittens', *Journal of Neurophysiology*, **26**, 996–1002

Levy, J. (1974), 'Psychobiological implications of bilateral asymmetry', in *Hemisphere Functions in the Human Brain*, eds Dimond, S. J. and Beaumont, J. pp. 121–83, London: Elek

Merrill, E. G. and Wall, P. D. (1972), 'Factors forming the edge of a receptive field: the presence of relatively ineffective afferent terminals', *Journal of Physiology*, **226**, 825–46

Patel, A. J., Balazs, R., Altman, J. and Anderson, W. J. (1975), 'Effect of x-irradiation on the biochemical maturation of rat cerebellum: postnatal cell formation', *Radiation Research*, **62**, 470–77

Purpura, D. P. (1975), 'Morphogenesis of visual cortex in the pre-term infant', in *Growth and Development of the Brain*, ed. Brazier, M. A. B., New York: Raven

Rabinowicz, T. (1977), 'The differentiate maturation of the human cerebral cortex', in *Human Growth*, Vol. 1, eds. Falkner, F. and Tanner, J. M., New York: Plenum

Rakic, P. (1976), 'Prenatal genesis of connections subserving ocular dominance in the Rhesus monkey', *Nature*, **261**, 467–71

Schultz, D. M., Gardiano, D. A. and Schultz, D. H. (1962), 'Weights of organs of fetuses and infants', *Archives of Pathology*, **74**, 244–

Tanner, J. M. (1977), *Education and Physical Growth*, 2nd edn, London: Hodder and Stoughton

Wiesel, T. N. and Hubel, D. H. (1963), 'Effects of visual deprivation on morphology and physiology of cells in the cat's lateral geniculate body', *Journal of Neurophysiology*, **26**, 978–93

Witelson, S. F. (1976), 'Sex and the single hemisphere: specialization of the right hemisphere for spatial processing', *Science*, **193**, 425–7

Witelson, S. F. (1977), 'Developmental dyslexia: two right hemispheres and none left', *Science,* **195**, 309–11

Yakovlev, P. I. and Lecours, A. R. (1967), 'The myelogenetic cycles of regional maturation of the brain', in *Regional Development of the Brain in Early Life*, ed. Minkowski, A., Oxford: Blackwell Scientific Publns

Chapter 9

Further Reading

Eveleth, P. B. and Tanner, J. M. (1976), *Worldwide Variation in Human Growth*, London: Cambridge University Press
A massive compilation of all known world data on children's growth made as part of the International Biological Programme in 1964–74. Though essentially a reference work, the text and figures deal quite succinctly with the environmental and genetic sources of differences between growth in different populations. Much of the documentation for the statements in this chapter is to be found here

Garn, S. M. and Clark, D. C. (1975), 'Nutrition, growth, development and maturation: findings from the ten-state nutrition survey of 1968–1970', *Pediatrics*, **56**, 306–19
A description of the most comprehensive attempt yet made to survey the nutritional status of the population of the United States. Documents exceptionally well the relation between income and growth rate

Garn, S. M. and Bailey, S. M. (1978), 'The genetics of maturational processes', in *Human Growth*, Vol. 2, eds. Falkner, F. and Tanner, J. M., New York: Plenum
A review of the genetic–environment interaction in respect of the timing of events such as the appearance of ossification centres, the eruption of teeth, and the events of sexual maturation

Goldstein, H. (1971), 'Factors influencing the height of seven-year-old children: results from the National Child Development Study', *Human Biology*, **43**, 92–111
A thorough analysis of the variation in heights of children in a truly national sample (all births in one week of 1958). Suitable only for readers whose statistical knowledge embraces elementary analysis of variance technique

Greulich, W. W. (1976), 'Some secular changes in the growth of American-born and native Japanese children', *American Journal of Physical Anthropology*, **45**, 553–68
A comparison not only of San Francisco-born and -dwelling with Japan-born and -dwelling Japanese, but of the relative secular trends in the two groups, showing how the Japan-born have caught up the California-born, in whom the trend stopped in the 1950s

Hiernaux, J. (1968), 'Bodily shape differentiation of ethnic groups

and of the sexes through growth', *Human Biology*, **40**, 44–62
An important paper both as to method and result. Shows how to follow the development of shape during growth, and how misleading indices such as sitting height/stature may be

Lindgren, G. (1976), 'Height, weight and menarche in Swedish urban school children in relation to socioeconomic and regional factors', *Annals of Human Biology*, **3**, 510–28
The first paper to report absence of effects of social class on height growth. A landmark in growth as a measure of the classlessness of society

Meredith, H. V. (1976), 'Findings from Asia, Australia, Europe and North America on secular change in mean height of children, youths, and young adults', *American Journal of Physical Anthropology*, **44**, 315–26
A recent review of the secular change in height, paying particular attention to the data on conscripted soldiers and stressing the part played by earlier completion of height

Richardson, S. A. (1975), 'Physical growth of Jamaican school children who were severely malnourished before 2 years of age', *Journal of Biosocial Science*, **7**, 445–62
Besides reporting a most careful study this paper reviews critically the results of previous malnutrition studies and discusses research designs and the generalizations which may currently be made. For the most part non-technical and highly recommended

Schreider, E. (1964), 'Recherches sur la stratification sociale des caractères biologiques', *Biotypology*, **25**, 105–35
A comprehensive review of relations between height, body build, mental ability and social class, in the UK and France

Stein, Z., Susser, M., Saenger, G. and Marolla, F. (1975), *Famine and Human Development: The Dutch Hunger Winter of 1944–1945*, New York: Oxford University Press
A very important (and very readable) account of the follow-up as adults of persons who experienced a brief period of severe famine as foetuses or infants and were subsequently fed at the usual Western–industrial level. The results emphasize the great resilience of the children who survived

Tanner, J. M. (1966), 'Galtonian eugenics and the study of growth. The relation of body size, intelligence test score and social circumstances in children and adults, *Eugenics Review*, **58**, 122–35, Reprinted in: *Trends and Issues in Developmental Psychology*, eds. Mussen, P. H., Largen, J. and Covington, M., New York: Holt, 1969)
A review of data from the early studies in the 1890s down to 1966. Discusses in particular the effect of social mobility on distribution of height

Tanner, J. M. (1968), 'Earlier maturation in man', *Scientific American*, **218**, 21–7

A review of data bearing on the trend towards earlier maturation and greater height in the industrialized countries

Thoday, J. M. (1965), 'Geneticism and environmentalism', in *Biological Aspects of Social Problems*, eds. Meade, J. E. and Parker, A. S., London: Oliver & Boyd

A sharp and accurate statement of the interaction of genetic and environmental influences in the moulding of phenotypic characteristics. Full of wisdom; the best introduction for the non-biologist to the nature–nurture problem

Waterlow, J. C. and Payne, P. R. (1975), 'The protein gap', *Nature*, **258**, 113–17

A most salutary review of the present situation in the world regarding infant malnutrition. The authors discount the notion that the lack is of protein and contend it is of total food, or energy. A good introduction to nutritional energetics

van Wieringen, J. C. (1978), 'Secular growth changes', in *Human Growth*, Vol. 1, eds. Falkner, F. and Tanner, J. M., New York: Plenum

A detailed review of the subject, with presentation of the very full Dutch records for height of conscripts during the last hundred years. Emphasizes the detailed relationships between environmental factors and changes in heights and weights

References

Babson, S. G. and Phillips, D. S. (1973), 'Growth and development of twins dissimilar in size at birth', *New England Journal of Medicine,* **289**, 937–40

Chang, K. S. F. (1970), personal communication

Cheek, D. B., Holt, A. B. *et al.* (1976), 'Nutritional studies in the pregnant rhesus monkey: the effect of protein-calorie or protein deprivation on growth of the fetal brain', *American Journal of Clinical Nutrition*, **29**, 1149–57

Cliquet, R. L. (1968), 'Social mobility and the anthropological structure of populations', *Human Biology*, **40**, 17–43

Damon, A. (1968), 'Secular trend in height and weight within Old American families at Harvard 1870–1965. 1. Within twelve four-generation families', *American Journal of Physical Anthropology*, **29**, 45–50

Elias, M. F. and Samonds, K. W. (1977), 'Protein and calorie malnutrition in infant cebus monkeys. Growth and behavioral development during deprivation and rehabilitation', *American Journal of Clinical Nutrition*, **30**, 355–66

Fleagle, J. G., Samonds, K. W. and Hegsted, D. M. (1975),

'Physical growth of cebus monkeys, *Cebus albifrons*, during protein or calorie deficiency', *American Journal of Clinical Nutrition*, **28**, 246–53

Fomon, S. J., Thomas, L. N., Filler, L. J., Zielger, E. E. and Leonard, M. T. (1971), 'Food consumption and growth of normal infants fed milk-based formulas', *Acta paediatrica scandinavica*, **223**, supplement

Friend, G. E. and Bransby, E. R. (1947), 'Physique and growth of schoolboys', *Lancet*, **2**, 677

Frisancho, A. R., Cole, P. E. and Klayman, J. E. (1977), 'Greater contribution to secular trend among offspring of short parents', *Human Biology*, **49**, 51–60

Frisancho, A. R., Sanchez, J., Pollardel, D. and Yanez, L. (1973), 'Adaptive significance of small body size under poor socioeconomic conditions in Southern Peru', *American Journal of Physical Anthropology*, **39**, 255–62

Goldstein, H. and Peckham, C. (1976), 'Birthweight, gestation, neonatal mortality and child development', in *The Biology of Fetal Growth*, eds. Roberts, D. F. and Thomson, A. M., *Symp. Soc. Study Hum. Biol.*, **15**, 81–108

Hamill, P. V., Johnston, F. E. and Lemeshow, S. (1972), *Height and Weight of Children: Socio-economic Status: United States*, U.S. Dept. of Health, Education and Welfare, Publn No. 73–1601, Rockville, Md

Hamill, P. V., Johnston, F. E. and Lemeshow, S. (1973), *Body Weight, Stature and Sitting Height: White and Negro Youths 12–17 years*, U.S. Dept. of Health, Education and Welfare, Publn No. HRA 74–1608, Rockville, Md

Hill, E. E., Mayers, R. E., Holt, A. B., Scott, R. E. and Cheek, D. B. (1971), 'Fetal growth retardation produced by experimental placental insufficiency in the rhesus monkey. 2. Chemical composition of the brain, liver, muscle and carcass', *Biology of the Neonate*, **19**, 68–82

Janes, M. D. (1975), 'Physical and psychological growth and development', *Environmental Child Health*, **121**, 26–30

Johnston, F. E., Hamill, P. V. V. and Lemeshow, S. (1974), 'Skinfold thickness in a national probability sample of U.S. males and females aged 6 through 17 years', *American Journal of Physical Anthropology*, **40**, 321–4

Jones, D. L., Hemphill, W. and Meyers, E. S. A. (1973), *Height, weight and other physical characteristics of New South Wales chil-*

dren. 1. *Children aged 5 years and over*, Sydney: New South Wales: Dept of Health

Kano, K. and Chung, C. S. (1975), 'Do American-born Japanese children still grow faster than native Japanese?', *American Journal of Physical Anthropology*, **43**, 187–94

Kerr, G. R., el Lozy, M. and Scheffler, G. (1975), 'Malnutrition studies in *Macaca mulatta*. IV. Energy and protein consumption during growth failure and "catch-up" growth', *American Journal of Clinical Nutrition*, **28**, 1364–76

Kiil, V. (1939), 'Stature and growth of Norwegian men during the last 200 years', *Skrifte Norske Videnskap Akademie*, No. 6, 175 pp.

Lamb, H. H. (1973), 'Whither climate now?', *Nature,* **244**, 395–97

Lindgren, G. (1976), 'Height, weight and menarche in Swedish urban schoolchildren', loc. cit.

Ljung, B. O., Bergsten-Brucefors, A. and Lindgren, G. (1974), 'The secular trend in physical growth in Sweden', *Annals of Human Biology*, **1**, 245–56

MacMahon, B. (1973), *Age at Menarche, United States*, U.S. Dept. of Health, Education and Welfare, Publn No. HRA 74–1615, Rockville, Md

Malcolm, L. A. (1971), *Growth and Development in New Guinea*, Madang: Institute of Human Biology (monograph No. 1)

Malina, R. M. (1973), 'Biological substrata', in *Comparative Study of Blacks and Whites in the United States*, eds. Miller, H. S. and Deeger, R. M., New York: Seminar Press

Marshall, W. A. (1971), 'Evaluation of growth rate in height over periods of less than a year', *Archives of Disease in Childhood*, **46**, 414–20

Marshall, W. A. and Swan, A. V. (1971), 'Seasonal variation in growth rates of normal and blind children', *Human Biology*, **43**, 502–16

Mueller, W. H. (1976), 'Parent–child correlations for stature and weight among school-aged children: A review of twenty-four studies', *Human Biology*, **48**, 379–97

Oduntan, S. O., Ayeni, O. and Kale, O. O. (1976), 'The age of menarche in Nigerian girls', *Annals of Human Biology*, **3**, 269–74

Panek, S. and Piesecki, E. (1971), 'Nowa Huta: Integration of the population in the light of anthropological data', *Materialy i Prace Antropologiczne*, **80**, 1–249

Peckham, C., Butler, N. and Frew, R. (1977), 'Medical and social aspects of children with educational problems', (in press)

Richardson, S. A. (1975), ed. cit.

Roberts, D. F. (1969), 'Race, genetics and growth', *Journal of Biosocial Science*, **1**, 43–67

Roberts, D. F., Danskin, M. J. and Chinn, S. (1975), 'Menarcheal age in Northumberland', *Acta paediatrica scandinavica*, **64**, 845–52

Rutter, M., Tizard, J. and Whitmore, K. eds. (1970), *Education, Health and Behaviour*, London: Longman

Shields, J. (1962), *Monozygotic Twins Brought Up Apart and Brought Up Together*, London: O.U.P.

Smith, D. W., Truog, W., Rogers, J. E., Greitzler, L. J., Skinner, A. L., McCann, J. J. and Harvey, M. A. S. (1976), 'Shifting linear growth during infancy: illustration of genetic factors in growth from fetal life through infancy', *Journal of Paediatrics*, **89**, 225–30

Stini, W. A. (1972), 'Malnutrition, body size and proportion', *Ecology of Food and Nutrition*, **1**, 121

Susanne, C. (1975), 'Genetic and environmental influence on morphological characteristics', *Annals of Human Biology*, **2**, 279–88

Tanner, J. M. (1962), *Growth at Adolescence*, 2nd edn, Oxford: Blackwell Scientific Publns

Tanner, J. M. (1964), *The Physique of the Olympic Athlete*, London: Allen & Unwin; distributed by Institute of Child Health, 30 Guilford Street, London w c1

Tanner, J. M. (1973), 'Trend towards earlier menarche in London, Oslo, Copenhagen, the Netherlands and Hungary', *Nature*, **243**, 95–6

Thomson, A. M. (1959), 'Maternal stature and reproductive efficiency', *Eugenics Review*, **51**, 157–62

Valman, H. B. (1974), 'Intelligence after malnutrition caused by neonatal resection of ileum', *Lancet*, **1**, 425–7

Widdowson, E. M. (1951), 'Mental contentment and physical growth', *Lancet*, **1**, 1316–18

Wilson, R. S. (1976), 'Concordance in physical growth for monozygotic and dizygotic twins', *Annals of Human Biology*, **3**, 1–10

Winick, M., Rosso, P. and Waterlow, J. (1970), 'Cellular growth of cerebrum, cerebellum and brain stem in normal and marasmic children', *Experimental Neurology*, **26**, 393

Yanagisowa, S. and Kondo, S. (1973), 'Modernisation of physical features of the Japanese with special reference to leg length and head form', *Journal of Human Ergology*, **2**, 97–108

Chapter 10

Further Reading

Hubel, D. H. and Wiesel, T. N. (1970), 'The period of susceptibility
    to the physiological effects of unilateral eye closure in kittens',
    Journal of Physiology, **206**, 419–36

Describes very clearly waxing and waning of the susceptibility between 4
and 12 weeks after the kitten's birth, and relates this to the development of
cells in the brain

Prader, A., Tanner, J. M. and Von Harnack, G. A. (1963), 'Catch-
    up growth following illness or starvation', *Journal of Paediatrics*,
    62, 646–59

The original description and designation of catch-up growth. Simple, and
requires no special medical knowledge

Scott, J. P., Stewert, J. M., De Ghett, V. J. (1974), 'Critical periods
    in the organization of systems', *Developmental Psychobiology*, **7**,
    489–513

A general review of the 'sensitive period' theory, with helpful diagrams.
Examples are given from socialization in the puppy

Tanner, J. M. (1963), 'The regulation of human growth', *Child
    Development*, **34**, 817–47. (See also *Nature*, **199**, 845–50,
    'Regulation of growth in size of mammals'.)

Describes a tentative theory of how the regulation of postnatal growth in
man works. Suggests that the organism is able to monitor both its actual
and its 'target' size and grow in accordance with the mismatch between
them

References

Jacobson, M. (1970), *Developmental Neurobiology*, New York:
    Holt, Rhinehart & Winston

Lorenz, K. (1960), *Discussions in Child Development*, Vol. 4, p. 128.
    Eds. Tanner, J. M. and Inhelder, R., London: Tavistock
    Publications

McGraw, M. (1943), *The Neuromuscular Maturation of the Human
    Infant*, New York: Columbia University Press

Money, J. and Hampson, J. G. (1955), 'Idiopathic sexual precocity
    in the male', *Psychosomatic Medicine*, **17**, 1–15

Tanner, J. M. (1962), *Growth at Adolescence*, 2nd edition, Oxford:
    Blackwell Scientific Publns

Tanner, J. M. (1975), 'Towards complete success in the treatment
    of growth hormone deficiency: A plea for earlier ascertainment',
    Health Trends, **7**, 61–5

Waddington, C. H. (1957), *The Strategy of the Genes*, London:
    Allen & Unwin

Chapter 11

Further Reading

Bayley, N. and Pinneau, S. R. (1952), 'Tables for predicting adult height from skeletal age: revised for use with the Greulich–Pyle hand standards', *Journal of Paediatrics*, **40**, 423–41, and **41**, 371
Describes the system of height prediction based on the Greulich–Pyle method for bone age

Hamill, P. V. V., Drizd, T. A., Johnson, C. L., Reed, R. R. and Roche, A. F. (1976), 'NCHS growth charts, 1976', *Vital Statistics Report*, H R A 76 – 1120, Vol. 25, No. 3, supplement
A general description of the U.S.A. cross-sectional growth charts, including height and weight for age; and weight for height, pre-puberty ages pooled

Healy, M. J. R. (1974), 'Notes on the statistics of growth standards', *Annals of Human Biology*, **1**, 41–6
A very simply written account of the use and meaning of standards for growth; the best introduction to the subject

Tanner, J. M., Whitehouse, R. H. and Takaishi, M. (1966), 'Standards from birth to maturity for height, weight, height velocity, weight velocity: British children, 1965', *Archives of Disease in Childhood*, **41**, 454–71; 613–35
The basic paper on growth standards, in which longitudinal standards based on both longitudinal data (for shape) and cross-sectional data (for amplitude) were introduced. Not light reading, but quite understandable to readers with an elementary knowledge of statistics

Tanner, J. M., Whitehouse, R. H., Marshall, W. A., Healy, M. J. R. and Goldstein, H. (1975), *Assessment of Skeletal Maturity and Prediction of Adult Height*, London: Academic Press
Gives illustrations of how bone age is assessed, and tables for predicting adult height. Before age 6 in girls and 8 in boys the tables do not involve bone age; at later ages an approximate prediction may be obtained, if bone age is unavailable, by assuming it to be average (i.e. equal to chronological age)

Waterlow, J. C., Buzina, R., Keller, W., Lane, J. M., Nickaman, M. Z. and Tanner, J. M. (1978), 'The presentation and use of height and weight data for comparing the nutritional status of groups of children under the age of 10', *Bulletin of the World Health Organisation* (in press)
The U.S.A. National Centre for Health Statistics charts are presented in this paper, which also contains a very clear discussion of the role of anthropometric measurements in nutritional surveillance

References

Hamill, P. V. V., Johnston, F. E. and Lemeshow, S. (1973), *Height and Weight of Youths 12–17 Years: United States*, U.S. Dept. of Health, Education and Welfare, Publn No. HSM 73–1606, Rockville, Md

Rona, R. J. and Altman, D. G. (1977), 'The National Study of Health and Growth. Standards of attained height, weight and triceps skinfold in English children 5–11 years old', *Annals of Human Biology*, **4** (in press)

Tanner, J. M. (1959), 'Boas' contributions to knowledge of human growth and form', in *The Anthropology of Franz Boas*, ed. Goldschmidt, W., Memoir No. 89 of the American Anthropological Association, Vol. 61, No. 5, Part 2, San Francisco: Chandler

Tanner, J. M. and Whitehouse, R. H. (1976), 'Clinical longitudinal standards for height, weight, height velocity and weight velocity and the stages of puberty', *Archives of Disease in Childhood*, **51**, 170–9

Chapter 12

Further Reading

Lacey, K. A. and Parkin, J. M. (1974), 'The normal short child. A community study in Newcastle-upon-Tyne', *Archives of Disease in Childhood*, **49**, 417–24
A very valuable survey of all children born in Newcastle in 1960 and below the 3rd centile for height at age 10, without any known specific disease as a cause. In most the shortness was familial, in some due to delay in growth, and in a few to psychosocial stress

Smith, D. (1976), *Recognisable patterns of congenital malformation*, 2nd edn, Philadelphia: Saunders
The medical textbook authority on the rare diseases causing short stature allied with various malformations and deformities. Not to be lightly consulted by the lay parent, for the illustrations may be upsetting; but a bible to paediatricians

Smith, D. W. (1977), *Growth and its Disorders*, Philadelphia: Saunders
An elementary textbook on disorders of growth. Has good reproductions of the standard growth charts. Written for medical students, but not upsetting to the lay parent

Tanner, J. M. and Thomson, A. M. (1970), 'Standards for birthweight at gestation periods from 32 to 42 weeks, allowing for

maternal height and weight', in *Archives of Disease in Childhood*,
45, 566–9
Standards for weight at birth, divided according to sex, length of gestation
and whether first- or later-born. The data are from Aberdeen, Scotland

References

Raben, M. S. (1958), 'Treatment of a pituitary dwarf with human
growth hormone', *Journal of Clinical Endocrinology and
Metabolism*, **18**, 901–3
Tanner, J. M., Lejarraga, H. and Cameron, N. (1975), 'The natural
history of the Silver–Russell syndrome: a longitudinal study',
Paediatric Research, **9**, 611–23

Author Index

244 *Index*

Subject Index

Where subjects appear *passim* references are given for definitive and selected entries only. Figures in bold type indicate definitions.

gr. = growth   sex diff. = sex differences

abnormal growth, 21
achondroplasia, 215
adolescence, **60**
adrenal glands, **95–7**
  disease of, 211
  see also under hormones
adults: gr. of, 19
  height prediction, *see* parents
age, postfertilization, postmenstrual, prenatal, **38–9**
alcohol and small-for-dates, 47, 211
amino acids, 28, 113
  peptides, **87**
androgyny, measure of, **70**
animal experiments, necessity of, 11
apes, cf. human gr., 22–3
  sex diff., 70
arm: motor development of, 108–10
  nerve impulses for age 2 yrs, 51
Assocn for Research in Restricted Growth (ARRG), 215, 217
athletic ability: muscle gr., 32
  gr. spurt, 76–7
axon, **105–6**

biochemical tests for foetuses, 47
birth, effect of, 49–51
birthweight: environment, nutrition and, 43–4
  first born, of, 44
  influence of maternal history, 43–4
  low, 44
blastocyst, **37**
blind children and seasonal gr., 142
boarding schools and gr., 102, 144
bone: age, **81**, 170–1
    prediction of menarche, 102
    standards, 201–2
  calcium and, 25–6
  cells, 24
  gr., 32–6
  primary centre of ossification, **32**
  skeletal ossification and sex diff., 58
boy, 'typical', **13**; *see also* sex difference

brain: acoustic analyser, **111**
  cerebellum, **105**
  glial cells, 106–7
  gr. curve, 15
  gr. spurt, 16, 70
  hypothalamus, **57**, 88, 89, 91
    and sex diff., 56–7
  injury at birth, 50
  maturation and evolution, 23
  motor area, **107**
  myelination, **110–11**
  receptor molecules, **87–8** ff
  reticular formation, **111**
  sensors, **91**
  sensory area, **107**
  sex diff., 56–7
  visual area, 107, 110–11
  visual system (experiments), 114–15, 118, 161
  see also neuroglia, neurons
breast milk and prolactin, 99
breech birth and GH deficiency, 213–14·
breathing at birth, 50

calcium turnover, 25–6
canalization and catch-up, **154–60**
cardiovascular system at birth, 50
cartilage: cells, 32–3
  response to hormones, 70
  disorders, 215
cat, visual system of, 115, 118
catch-up gr., **155**, 154–60
  after birth, 42
  hypothyroidism, 159
  malnutrition and starvation, 128–35, 155
  miniatures, and, 91–2
  'target' theory of, 159–60
cells: brain, gr. in, 106 ff
  competence, 160–1
  differentiation (morphogenesis), 39
  membrane and nucleus, **26**
  regeneration of, 24–5, 29–31
  specification, 160–1
  structure, **26–9**

Inst. of Biological Anthropology
58 Banbury Road, Oxford. OX2 6QS

Tel: 0865 - 274700   Fax. 0865 - 274699